Knowings

In the Arts of Metaphysics, Cosmology, and the Spiritual Path

Charles Upton

KNOWINGS

In the Arts of Metaphysics, Cosmology, and the Spiritual Path

SOPHIA PERENNIS

SAN RAFAEL, CA

First published in the USA
by Sophia Perennis
© Charles Upton 2008

Series editor: James R. Wetmore

For information, address:
Sophia Perennis, P.O. Box 151011
San Rafael, CA 94915
sophiaperennis.com

Library of Congress Cataloging-in-Publication Data

Upton, Charles, 1948–
Knowings : in the arts of metaphysics, cosmology,
and the spiritual path / Charles Upton.

p. cm.
ISBN 1 59731 074 1 (pbk: alk. paper)
1. Metaphysics. 2. Cosmology. 3. Spiritual life. I. Title.
BD113.U68 2008
110—dc22   2008003155

CONTENTS

Preface

Metaphysics is the study of eternal first principles, and of God as the First Principle of all, both Pure Being—the personal God to Whom we pray, and Who necessarily presents to us a personal face, since we ourselves are persons—and the Formless Absolute, beyond even Being itself, out of which the personal God eternally arises into Being, and into which He eternally returns, in a single motionless act. *Cosmology* is the study of the universe conceived of as created by God—and also, in a larger sense, as the bursting into dimensional manifestation of the Formless Absolute. And the spiritual path is the way by which the human soul, understood as the *microcosmos*, the epitome of universal manifestation, is returned to its own essential nature, and conformed to Absolute Truth.

The sense that there is such a thing as Absolute Reality that is forever beyond not only the senses and the world they report, but also the worlds of thought, feeling and imagination that the faculties of our soul perceive—eternally beyond them, but also rhythmically manifested by them—is the origin of what is called 'the sense of the sacred'. When we lose this sense, we inevitably forget not only what Reality is, but what a human being is—and this, since we are in fact human beings, is the ultimate tragedy. Human dignity, the recognition of natural rights, the sense that life has a meaning, a purpose and an ultimate destiny, all these disappear—eventually, or immediately—when the metaphysical instinct is lost. Metaphysics, cosmology and the spiritual path are sciences in the sense that the principles they are based upon are objective and universal, but they are also arts. In the concrete practice of metaphysics that constitutes the spiritual path, universal principles confront diverse situations and unique individuals; thus the 'repeatable experiments' of science must give way at one point to the intuitive touch of the artist. The one who truly masters any science transforms that science into an art.

Metaphysics is the intellectual guardian of true religion, just as religion is the existential proof of true metaphysics: to know God is to love Him, and to love Him is to receive, by His grace, the will and the power to actualize Him, down to the marrow of our bones.

This is the ultimate purpose of human life.

† † †

There is one passage in this book that appears in two different essays: 'Christianity, Hinduism and Islam: *A Dialogue with Robert Bolton*', and 'Christian and Muslim "Trinitarianism": *A Reply to Philip Sherrard*; this is done deliberately, in view of the fact that the same insights have different resonances when placed in different contexts. Certain passages in the dialogue are repeated as well, to help the reader follow the ins and outs of the argument.

The names René Guénon and Frithjof Schuon that appear in many places throughout this book may not be familiar to some readers. René Guénon is often thought as one of the two founders—along with Ananda Kentish Coomaraswamy—of what has come to be called the 'Traditionalist' or 'Perennialist' school. Writers associated with this school include Martin Lings, Titus Burckhardt, Marco Pallis, Julius Evola (who is usually excluded by those Traditionalists who follow Frithjof Schuon), Charles le Gai Eaton, Rama Coomaraswamy, Whitall Perry, Mark Perry, Elemire Zolla, Tage Lindbom, Lord Northbourne, Joseph Epes Brown, Philip Sherrard, Seyyed Hossein Nasr, Wolfgang Smith, William Chittick, Jennifer Doane Upton, Patrick Laude, Jean-Baptiste Aymard, Reza Shah-kazemi, and James Cutsinger, among others (including, more tangentially, such scholars as Henry Corbin, Louis Massignon, Mircea Eliade and Huston Smith). Between them, these writers have brought comparative religion to perhaps its highest historical development, and have explicated and represented almost the entire human store of traditional metaphysics and esoterism, with a view to the needs and challenges of our time. The three greatest names on this list are undoubtedly René Guénon, who almost single-handedly resurrected 'pure' metaphysics for the western world and brought many of the traditional doctrines of the Vedanta and Taoism to the west for the first time (at least in unadulterated form), as well as producing a telling critique of many types of false and dangerous occultism that distort the traditional metaphysics of both east and west; Ananda Coomaraswamy, the great expositor of traditional art theory and Hindu metaphysics, who was equally conversant with the philosophy, theology and metaphysics of the western world; and Frithjof Schuon, who brought metaphysical exposition and comparative religion to their highest level of development in the modern west, and in so doing enunciated (along with Guénon) the essential principles by which traditional spiritualities much be approached in this most anti-traditional of times.

PART ONE

Metaphysics and its Echoes

Reality Is

A Response to Postmodernism

If the world our senses report were perfectly objective and absolutely real, then it would be God. But it isn't—so it's not. Because every point-of-view or way of viewing that world is different, perfect objectivity cannot come by the senses.

If our consciousness were perfectly subjective and absolutely real, never impinged upon or in need of the outer world of the senses, then it would be God. But it isn't—so it's not. Perfect subjectivity cannot come by psychic introspection.

Nor can absolute reality be attributed to the mysterious interpenetration of the objective and subjective worlds. Taking the sense-world, including the human brain, as materially real, and the basic substance of things, consciousness can only be defined as an epiphenomenon of matter. Taking consciousness, including the experience of an outer world, as real, and the basic substance of things, matter can only be defined as a modification of consciousness. But if, from the standpoint of the primacy of consciousness, matter can only be defined in terms of consciousness—and if, from the standpoint of the primacy of matter, consciousness can only be defined as a product of matter—and since no third point-of-view outside the polarity of matter and consciousness can be delineated, because any such point-of-view would simply be either one more product of matter or one more state of consciousness—then neither pole can be absolutely real, nor can their interrelationship be absolutely real, given that neither pole can be defined without recourse to the other, in view of the fact that matter and consciousness are never found apart. If we are forced to say both that matter only

exists because I am conscious of it, and that I am only conscious
of anything because matter exists, then neither can matter be
the sufficient definition of consciousness nor can consciousness
be the sufficient definition of matter. Consciousness without
matter cannot perfectly define itself because it can never be
totally free from its own subjectivity, never stand outside itself
to see, objectively, what it is, unbiased by 'consciousness'—nor
can matter without consciousness even begin to define itself,
because 'defining' is necessarily a conscious act.

To repeat: Neither the world reported by the senses nor the
world of consciousness can be absolutely real, nor can the either
term be only an epiphenomenon or a reflection of its opposite.
Both, however, are host to the signs of an Absolute Reality
which transcends them.

The reality of the Sun is not determined by my consciousness
of it. The Sun is really out there. True, no two measurements of
the distance between the Earth and the Sun can be absolutely
identical. On the other hand, every measurement of that distance
approximates to all the other measurements. As instruments and
techniques are refined, the measurement—not just for me, but
for every trained scientist upon Earth—becomes more accurate.

So the differences in our various subjective experiences of the
objective world in no way disprove the existence of such a
world, since all such experiences *approximate* to total objectivity,
without, however, ever finally reaching it. Sense experience,
then, does not directly reflect total objectivity, but rather 'trian-
gulates' it. It does not perfectly report the reality of an objective
world, but it does prove that such a world really exists. Phenomena approximate to, and thus 'triangulate', the Noumenon.

Nor does the existence of an objective world beyond our subjective experiences of it, which can never perfectly reflect it,
prove our subjective experiences to be illusory, or less real than
that objective world, mere epiphenomena of it. The sense that 'I'
exist, apart from any other content of consciousness, is universal
to all perceiving subjects, and is in no way fundamentally altered
(though it may exist at greater or lesser degrees of intensity) by
the multiplicity of perceiving subjects. To paraphrase Shankara-

charya, 'the feeling that "I Am" is a sign of the reality of the *Atman*, the Absolute Self within me.' Just as the multiplicity of ways of perceiving the objective, outer world does not disprove the existence of that world, so the fact that many relative and limited subjects all experience the fact that 'I Am', does not disprove the existence of the *Atman*. As multiple instances of sense experience triangulate the Noumenon, so multiple instances of self-consciousness triangulate the Self.

The Noumenon behind the screen of the many subjectively-modified experiences of the objective sense-world, and the Self behind the veil of the many subjectively-modified experiences of self-consciousness, are the same Reality, which exists 'prior' to both experience and things-to-be-experienced, prior to the polarization between subject and object which creates the 'universe'. The world is not my creator; neither is it my fantasy. In the words of Frithjof Schuon, 'The world may be a dream, but it is not my dream.' It is none other than the Self-manifestation of the Absolute Reality via polarization into subject and object— into many limited self-consciousnesses enclosed within an imperfectly-perceived world.

Only the existence of God, only the existence of an Absolute Reality Who transcends the polarity between consciousness and matter, subject and object, can solve the philosophical contradictions which inevitably arise when we try to take consciousness as absolute, as in the case of idealism, or matter as absolute, as in the case of materialism. The evidence of the existence of an objective, outer world is undeniable but never perfect; only God is absolutely objective. The evidence for our own subjective reality via the sense that 'I Am' is undeniable, but never perfect; only God is the perfect and self-sufficient Self.

The Absolute Self is the Absolute Objectivity behind the phenomena of the world; the Absolutely Real Object is the Absolute Self behind our contingent subjectivities. What in cosmic dimensionality is polarized into subject and object, in God is perfectly One. In the words of the Qur'an: *I will show them My signs on the horizons and in their own souls until they are satisfied that this is the Truth. Is it not enough for you that I am Witness over all things?*

If we no longer believe in God as an Absolute Reality transcending subject and object, we fall into the hopeless philosophical contradictions of trying to define subject in terms of object,
or object in terms of subject, or both in terms of one or the
other, or both in terms of both. And these philosophical contradictions have real, and dire, human consequences. If the world is
only my fantasy, then other people are not real—either that, or
the only thing I can ever see of them is my own construction of
them—my own pattern of experience—my isolated ego-self.
Such philosophical narcissism, being the basis of emotional narcissism, is the death of love. And if I am not real, if I am nothing
but a modification and epiphenomenon of a material world, if I
have no soul, then self-respect and human dignity become
impossible—in the case of other people as well as in my own
case. If I have no basis for respecting myself, I certainly have no
basis for respecting others—just as, if others are not real to me, I
have simultaneously lost the prerogative of calling myself real.

God is the only bridge between subject and object, since He
the One Reality of which they are polar modifications. Therefore only God allows me both to credit the reality of the other—
to see her and treat her as a person in her own right, as something more than simply a projection of my own beliefs, fears and
desires—and to realize that, since she is not merely an object of
my consciousness but a subject in her own right, then I am, at
least potentially, as real to her she is to me. The Absolute Reality
of God allows me to intuit my own integral reality in His Sight, a
reality which I am then compelled to grant to all the others who
also stand in that Sight. In other words, God is the end of both
self-alienation and of my habit of treating others only as objects
of my consciousness, these being the existential roots of co-
dependency, domination-and-submission, and all the other dis-
eases of human relatedness. In other words, God is Love.

Mythopoeia
and Metaphysics

Metaphysical truth can be told in several languages: The language of philosophical discourse; the language of music; the language of visual symbols—icons, statues, calligraphy, dance, gesture, symbolic designs, games, sacred architecture, esoteric horticulture in the form of symbolic gardens, traditional clothing, weapons, and tools; and mythopoeia, the language of poetry, fable, parable, myth, liturgy, folklore and folksong. Like the language of visual symbols, mythopoeia addresses itself to the deep stratum of the soul, unconscious to most of us most of the time but available via the dream-state, where intelligible realities first clothe themselves in symbolic images, and by so doing set intuitive and emotional waves in motion through the psyche, like the widening ripples on the surface of a pond when a stone is dropped. But because mythopoeia does not present images directly, but rather evokes them by means of language, it acts to bring together the two dominant modes of human understanding: the 'left-brain' mode mediated by words and the 'right-brain' mode catalyzed by images. When the unconscious mind is stocked and informed with a rich, traditional mythopoeia—with material that is truly *legitimate*, conformed to and emanating from the intelligibles or Platonic ideas, not spurious as in the case of idle or self-interested fantasy, or demonic illusion—it provides a fertile substratum that allows the conscious, *discursive* understanding of metaphysical principles to take root in the psyche, on a level deeper than thought. And such conscious understanding can also act as an *hermeneutic* for mythopoetic narrative and imagery, revealing to the conscious mind the metaphysical principles latent within mythopoeia, allowing spiritual Truth to permeate both thought and feeling, and by so doing reconcile them. And if the will is concurrently being taught how to conform itself to virtue, social and spiritual duty, and self-respect, these together can be said to comprise the alchemical *magnum opus*, the production of the Philosopher's Stone.

The quintessential vehicle of mythopoeia is poetry, which—at least in its most intense and concentrated expressions—can be numbered among the final reverberations within the soul of God's creative act. Poetry extends the divine creativity far and wide within the human psyche, both individual and collective; it carries that Truth out of which, according to the Holy Qur'an, all things are made, to its ultimate psychic limits—in other words, as far as to the threshold of unreality, evil and non-existence. This is the great danger of poetry, to both the poet and the society around him, and the reason why the practice of it, outside of a traditional liturgical context, carries inevitable spiritual perils—as witness the alcoholism, drug addiction and suicide of so many poets in modern times. Poetry is the language of the gods. The poet, however, is not a god but a man—a man who has, as it were, stolen the divine fire, the ability to create *icons*, living images of truth. If his skill is great enough, these icons will inevitably command belief—not in the form of assent to clear and true doctrine, but in terms of the kind of emotional and intuitive allegiance that only true doctrine deserves. Consequently, if the iconic forms wrought by a poet are not objectively true as well as subjectively convincing, he has arrogated to himself the godlike power to *determine what is true by saying it,* and perverted that power. Only God can say what is to be true; if a poet attempts to do so outside of God's inspiration and permission, he has become what Plato, in the *Republic,* calls a 'liar'. And this is a form of demonic invocation. According to the Qur'an, in the surah 'The Poets':

> Shall I inform you upon whom the devils descend? They descend on every sinful, false one. They listen eagerly, but most of them are liars. As for the poets, the erring follow them. Hast thou not seen how they stray in every valley, and how they say that which they do not? Save those who believe and do good works, and

remember Allah much, and vindicate themselves after they have been wronged.

To *say* something but not *do* it is to extend the name and image of Reality into imaginative forms that one has neither the power, the integrity, or the *right* to realize. It is to create phantasms, to go into debt to Reality Itself, and thereby to wrong oneself, sometimes mortally. Poetry is boast, only action is proof; the poet who vindicates himself after having wronged himself is the one who has paid, with spiritual suffering and warfare, the debt he incurred when he arrogated to himself the divine power of creative speech.

And what is true for poetry is even more true for the expression of metaphysical ideas. In the spiritual Heart, knowing and being are one, but when the intuitive/thinking/expressive mind catches a reflected flash from the light of that Heart, and suddenly 'knows' more than it *is*, it has gone into debt to the Truth. For the one open to metaphysical intuitions, the payment of that debt—by that one's own dedicated efforts, and by the generous grace of God—is the essence of the spiritual Path.

Who is the Doer?

According to the Hindu Vedanta, God is not the Doer, but the Witness. Action arises within the relative world defined by the three *gunas* or modes of universal Substance (*Prakriti*, roughly analogous to *Maya*)—namely *sattva*, clarity and purity, the rising tendency; *rajas*, vigorous activity, the expanding tendency; and *tamas*, ignorance and obscurity, the sinking tendency. These qualities of the relative world—of which you and I, as human actors, are a part—produce all action by their mutual effects, while God, as *Purusha*, is the Witness of them, just as God as *Shiva*, the holder of power, is motionless and impassive, beyond all activity, while his *Shakti*, his emanation or feminine aspect, is dynamic, creative, and re-integrating. The highest and most comprehensive conception in Hinduism of God as Witness is the doctrine of the *Atman*, the universal presence of the Absolute Witness as the One Self of all. In the words of the Bhagavad-Gita:

> *Who sees all action*
> *Ever performed*
> *Alone by Prakriti,*
> *That μman sees truly:*
> *The Atman is actless.*

The opposite (or apparently opposite) conception can be discerned in the Abrahamic religions. Like the Hindu *Brahma*, the creator, or like the gods *Indra* or *Ishvara* insofar as they may be taken as different guises of the One as Governor and Administrator of the universe, God is most often conceived, in Judaism, Christianity and Islam, as pre-eminently active rather than contemplative. God is omnipotent, and one of the not-always-elaborated but nonetheless inevitable implications of His omnipotence is, not simply that He can do anything He desires, but that all action, whoever the apparent actor may be, ultimately flows from Him. If 'It is not I who live but Christ lives in me,' then it must also be true to say that 'It is not I who acts, but Christ acts in me.'

The most explicit expression of the doctrine that only God is the Doer is to be found in the Asharite school of Islamic *kalam*, which denies secondary causes (comparable in some ways to the *gunas*) and attributes everything—all forms, all events, even all *choices*—to the sovereign will of God, even going so far as to assert that events are not even proximately attributable to natural law, or secondary causes within the realm of creation, but are produced directly by God through His continuous re-creation of the entire universe, instant by instant—the famous doctrine of 'occasionalism'. Asharite doctrine would thus seem to deny free will, and did in fact produce, in Islam, a bias toward fatalism as against the voluntarism of the rationalist Mutazilite school, whom the Asharites ultimately defeated (though not without incorporating some of their doctrines and dialectical methods). It is nonetheless clear from the Qur'an that the absolute denial of free will is in no way Islamic. On almost every page of the Holy Book warnings and exhortations appear, which, if there were no such thing as human free will, would be meaningless. God does not overwhelm us, forcing us to do this or to choose that; rather, He seconds us in our own choices, guiding those who submit to Him and leading astray those who rebel against Him. In the words of the Qur'an (14:4), *Allah sendeth whom He will astray, and guideth whom He will. He is the Mighty, the Wise.* If we choose to follow Him, the knowledge and strength we require for His service are attributable only to Him, not to us, just as our very being is not attributable to us, but is freely given out of the store of His Being. That knowledge and strength are drawn from Him in his names *Al-Nur* (Light), *Al-Hadi* (the Guide), *Al-Muqit* (the Nourisher), and *Al-Mujib* (The Answerer of prayers). Likewise if we choose to disobey Him, the impulse to transgress is drawn only from Him, in his Names *Al-Jabbar* (The Compeller), *Al-Muqtadir* (the All-Determiner), *Al-Mudhill* (the Abaser), and *Al-Darr* (the Punisher); since God is omnipotent, our power to rebel against him is granted to us only by His leave. No choice made by man, free though every choice most certainly is, has the power to depart from the will of God, which embraces all actions and all choices. God does not reward or punish us in recompense for our actions, but by means of them.

From the Christian standpoint, it is God's omniscience, His foreknowledge of our choices, that seems to threaten the doctrine of free will; if God knows ahead of time whether we will sin or embrace virtue, whether we will be saved or eternally damned, then of what use is human action? The Christian philosopher Boethius dealt definitively with this objection, in his doctrine that since God dwells in the eternal present, He does not *foresee* our choices but rather *witnesses* us making them—and to witness someone performing an action is not to make him perform it. (In conceiving of God as the Witness of action rather than the Author of it, Boethius is actually quite close to the Vedanta.)

From the Islamic standpoint, on the other hand, it is God's omnipotence rather than His omniscience that seems to leave no room for free will. If all things are determined by God's will, then our choices are also determined; as with the Calvinist conception of predestination, God is in danger of being conceived as an arbitrary tyrant who saves or damns us for His own inscrutable reasons, leaving us no way even to follow His will on our own initiative, much less to oppose it. Among others, it was Ibn al-'Arabi, the *Shaykh al-Akbar*, the 'greatest Sufi shaykh', who definitively refuted this error. Ibn al-'Arabi accepted Asharite occasionalism, but he interpreted it on a much higher and deeper level than the Asharites themselves. The basic rationale for the Asharite doctrine that all things are determined by God's will was the imperious need to avoid ascribing any 'partners' to God, a doctrine that is entirely in line with the clear meaning of the Qur'an. The associating of partners with God is the cardinal sin in Islam, the sin of *shirk*, comparable in some ways to the Christian 'sin against the Holy Spirit'. If God is seen as only *partly* responsible for the creation and governance of the universe and the events that occur within it, with natural law and human action also lending a hand so as to make up some kind of divine deficit, then His Omnipotence is denied. This is clearly a heresy, as any similar conception certainly would be within a Christian framework—I am thinking of the deism of the Enlightenment, and also of the modern theological conception that defines man as a 'co-creator'. To assert the doctrine of free will is one thing; to look upon God as weak and in need of help from man, or from natural law, is quite another. To delegate power is not to abandon it, and if an omnipotent God is capable of ordaining events, then we are in no way justified in denying Him the power to ordain laws, as long as we also ascribe to Him the power to suspend those laws if He so chooses, thus producing what we term 'miracles'.

The Asharites, however, did not speak of delegated power, but only of the direct action of God universally applied. Ibn al-'Arabi's response to this doctrine was to accept it, but also to assert that 'the determined determines the Determiner.' God, by his omnipotence, determines all things, but He determines them according to their natures, not in opposition to them. Those natures spring, by God's creative will, the *nafas al-Rahman* or 'breath of the Merciful', from the *ayan al-thabita*, the 'perma-nent archetypes' of all things in God, given that He, as Necessary Being, must embrace Possible Being as well—not as a privation, however ('possibility' in the sense of 'the merely possible, not the actual') but as an expression of His almighty creative power (possibility in the sense of 'with God, all things are possible', implying that within the Divine Nature, within the embrace of Necessary Being, all possibilities are actualized). God may cer-tainly seem to contradict Himself from the limited human point of view, but in His Own Nature He does not contradict Himself. Self-contraction is a privation, a weakness—and God, since He is omnipotent, has no trace of privation or weakness in Him. Ibn al-'Arabi's doctrine here is similar in many ways to that of St. Thomas Aquinas, who (in Stratford Caldecott's paraphrase) asserted that 'while God is free—for example to create or not to create, to redeem or not to redeem—His decisions are neither determined nor arbitrary, but "fitting".'

Nonetheless, the Hindu doctrine that God is the Witness of action, not the Doer, and the Islamic doctrine that in the last analysis God is the *only* Doer, appear to be poles apart. And yet the spiritual effect of both doctrines is identical: to break our egoistic identification with our own power to act. To see God as the only Doer, the ultimate Agent of our actions, is to mortify our self-will, and place our reliance on Divine Providence. Likewise, to contemplatively view our own actions from the standpoint of the Atman, the Absolute Witness within us, as instances of the universal dance of Maya (which, if it does not absolutely determine all of our actions, has at least provided us with both the circumstances we face and the unique human character through which we must face them), is equally to place ourselves beyond the reach of self-will. Furthermore, from one standpoint, the Maya we witness in contemplation, from the vantage point of the God within us, is equally the Providence we submit to in faith, as the will of the God Who lies beyond us. The great Hindu sage Ramana Maharshi in fact taught that there are two ways to reach *moksha* or Liberation. The first is to ask 'who am I?', recognizing that ultimately one is neither the body, nor the psyche, nor even the basic sense that 'I am', since—from the vantage point of the Atman—it is possible to witness the I-sense itself arising and then passing away, this witnessing being the essential practice of *jñana-yoga*. And the second way, available to those who are not destined to be liberated by means of the first, is to perfectly submit to the will of God. If we submit to God, abandoning all self-will, we will perceive God as acting within us, through our own actions and states of consciousness, as well as upon us, through external events. As it says in the Qur'an (41:53) *We will show them Our signs on the horizons and in themselves, till it is clear to them that it is the truth. Suffice it not as to thy Lord, that he is witness over everything?* And to perceive God as *acting* in us is only one step away from understanding Him as *knowing* and *being* in us. The full realization of this truth unveils the Atman, the Eye of the Heart, through which all events are witnessed as acts of God, and God-as-Actor is Himself witnessed as objective to the Absolute Witness, the Godhead within.

Since Truth is One, there is ultimately no contradiction in the Ways. To realize God as the Absolute Witness is to witness God as the One and Only Doer; there is no separation.

On Hinduism, Christianity, Islam, and Non-Dualism

A Dialogue between Charles Upton & Robert Bolton

Dear Robert Bolton

As I see it, your idea of the Vedanta—though one might make the point that this is actually the way Guénon presents it—contains certain misconceptions which seem to be very common among Christians. They are:

1) That the Vedantic Absolute is strictly 'impersonal'.

2) That the Vedanta divides everything between an infinite God and an illusory and cosmos.

3) That the Impersonal Divinity must be an 'object', since the Personal Divinity can be nothing other than a 'subject'.

I

The first misconception, on the sentimental extreme of the spectrum, becomes the Chestertonian image of the Mysterious East as an abyss of numb impassivity and terrible, impersonal heartlessness. This simply indicates that a personalistic sentimentalism must see all that transcends its own level as a demonic emptiness, void of all life, love and relatedness—and this may indeed be the realm encountered by some westerners who have been attracted to the eastern religions because they are basically in flight from God. To them, an impersonal Absolute seems less threatening than a personal God who is WATCHING US for God's sake, and Who may even REQUIRE something of us. An impersonal Absolute seems much less inconvenient; as C.S. Lewis said about the God of pantheism, 'He is simply there if you need him, like a book on a shelf; there is no danger that heaven and earth will flee away at his touch.' I would hazard a generality that Christians, or those with a Christian cultural background, will tend—consciously or not—to view the Vedanta as if it as were a kind of Greco-Roman pantheism, which is certainly not the case. And this misconception will present itself equally to those attracted by the Vedanta and to those repelled by it. I'll deal with this misconception in greater depth under my number 3; here I only want to say that to us westerners the word 'impersonal' denotes something on a lower level of being than personhood, like 'the Force' in the Star Wars mythology, something on the order of electricity or magnetism or nuclear energy. But the 'impersonal' Absolute is actually TRANSPERSONAL, otherwise the Personal God could not be its first and highest intelligible manifestation. It (or He?) transcends what we know as personhood in the same sense—though to an infinitely greater degree—in which you or I, as persons, transcend a stone. To say that God is only or essentially personal MAY be to imply that He is no more than we conceive Him to be; it may be to imprison Him on our human level of understanding, to deny that He opens out 'behind' onto the Infinite. But of course we habitually do the same thing in our conceptions of other people, and ourselves; we treat others as if they were no more than our ideas of them, and ourselves as if we were limited to our self-images; we forget that ALL persons are, precisely, personal faces of the same Transpersonal Mystery, because they are made in the image and likeness of God. As an icon of Christ is not Christ Himself but a window opening onto His Reality–which is ultimately the reality of the Father Whom 'none has seen at any time', given that 'I and the Father are One'—so you and I are 'icons' of the Universal Humanity, as Paul indicated when he said, 'It is not I who live, but Christ [Who is One with the Father] lives in me.'

II

The Vedanta does not strictly divide reality between an Infinite God and an illusory cosmos. To begin with, Maya does NOT mean 'illusion;' it means 'manifestation' or 'magical apparition', deriving from the root 'to measure'; Maya, then, in the Infinite when seen according to any finite conceptual or perceptual set. God creates the universe by Maya-power, projecting it as something which exists in one sense, and in another sense does not. The classical metaphor for the action of Maya in the Vedanta is 'to mistake a rope for a snake'. The 'snake' is clearly an illusion; the 'rope' is not. Maya is a manifestation of the God Who is unknowable in His Essence. If we take the universe as something existing in its own right, as something which would continue to exist even if God were to withdraw His attention from it, then we are deluded by Maya. The universe does not exist in its own right; it is a creation of God, Who has not simply created it in the past, but holds it in existence in this moment. It is created ex nihilo in the sense that God creates the universe from nothing ELSE than Himself, since only He possesses Being in His own right; the universe does not. In one sense it is a manifestation of Him: 'the heavens show forth the glory of God, and the earth declares His handiwork.' In its own right, it is nothing. And if we believe that it exists in its own right, then Maya has deluded us; in this sense alone can Maya be translated as 'illusion'.

Furthermore, the Vedanta does NOT make a strict separation between God and cosmos. If no separation at all were made—if the level where such a separation applies were not recognized—then the Vedanta would indeed be pantheism. And, as you say, if the Vedanta were to absolutize this separation in a simple way, then it could in no way be called an esoterism. In reality, the Vedanta recognizes four levels of consciousness, which are equally four ontological (or trans-ontological) levels: 1) 'Brahman is real, the universe is unreal'; (2) 'There is only Brahman'; (3) 'I am Brahman'; (4) 'All this, too, is Brahman.' And the earlier levels are not negated by the latter, but rather contained by them. Thus level 4 is not pantheism, because it embraces level 1, which negates pantheism, nor is level 3 megalomania, because it embraces level 2, where the individual self does not appear, as well as level 1 where, though it appears, it is recognized as illusory.

To say (as you did in your original email to James Wetmore) 'the reality of the universe is like the image of the Sun reflected on the water' is not, in my opinion, simplistic; rather, it is SIMPLE in the sense of immediately efficacious and accessible. Any child could understand it—in a childish way—but how many of us can really SEE the world around us, and our own phenomenal selves, as direct reflections of the Absolute? We can only pray that a lifetime of spiritual purification will enable us to catch a glimpse of this level of Reality. Here we come to one of the great apparent divides between the Vedanta and what some would call 'exoteric' Christianity— or that between, say, Plotinus and Semitic monotheism: the seeming conflict between 'emanationism' and 'creationism'. For God to 'emanate' the universe rather than creating it, as in the case of the appearance of the image of the Sun in a motionless body of water, seems to make creation an 'automatic' reflex of the Divine Reality, and thus to constrain God by something less than He is, something that is merely on the order of natural law—as if God were helpless NOT to create the universe, and thus, in effect, helpless also to deliberately create it. On one level, we can say that whereas Beyond Being emanates the universe—if we can actually place Beyond Being in relation to its own emanation as 'other', which strictly speaking we cannot—the Personal God, or Pure Being, creates it. In other words, the Absolute (as it were) brings the universe into being by first 'emanating' Pure Being, the Creator. In Vedantic terms, the first 'reflex' of Brahman is Ishvara, who does indeed plan, create, govern and maintain the visible universe; even Ramana Maharshi asserts this—though he adds that, from the standpoint of *jñanic* realization, Ishvara is simply the 'last thought'. While we experience ourselves as actors, God is the Supreme Actor whose actions supersede ours; by our own actions we can create only certain modifications in the conditions of our lives, while God the Creator has established both the entire range of those conditions, and ourselves as acting subjects with free will. But as soon as we transcend the experience of ourselves as authors of our own actions—by means of the realization that, in reality, God is the only Actor—then (paradoxically) we have also transcended God as Actor and Creator, at which point all things are viewed not as objects created by Him, but rather as direct emanations or reflections of His essential nature.

III

To say that the Impersonal Absolute (presumably Nirguna Brahman or God-without attributes) is OBJECT, while the Personal God (Saguna Brahman) is SUBJECT, is not what the Vedanta teaches. It reaches precisely the reverse: that the Absolute Witness or Atman is, in Frithjof Schuon's phrase, 'the Absolute Subject of [or behind] our contingent subjectivities,' whereas the world of conditions, taken (on one level of consciousness) as Saguna Brahman, is OBJECTIVE to this Witness; Beyond Being is the Absolute Witness of Being and all It creates—with the understanding that it does not witness Being as OTHER than Itself, but rather AS Itself.

At this point we can come to a deeper understanding of the Vedantic Absolute, in the mode of Atman, not as impersonal, but as transpersonal. That in me which witnesses things is my very power of consciousness, my very Self, the furthest thing from anything impersonal. And yet that Self nowhere appears in the total field of the possible objects of consciousness, since anything I witness out there as 'myself' is not my true Self, but merely a self-image, or a sense-image of my body; (remembering Blake's doctrine, from 'The Marriage of Heaven and Hell', that 'the Body is the portion of the Soul perceived by the five senses'). Who I Really Am never appears, CANNOT EVER appear, as an object of consciousness; in Vedantic terms, 'the eye cannot see itself'. The very essence of my personhood is thus not IMPERSONAL, but rather TRANSPERSONAL. What could be more obvious than this? And what is more hidden, to our passion-darkened habitual consciousness, than the obvious? It is not I who see the world, and the self I think I am, but Christ who sees it through my eyes. If

I seek to retain my soul, the self I think I am, I will lose it. But if I lose it for His sake, I will find it. Three (or four) levels of consciousness are described here. The first is the level of 'seeking to keep our souls', the level of our habitual egotism where we, in effect, believe that we have created ourselves, or at least that it is up to us to define ourselves, and to maintain our identities as so defined. (If we cannot transcend this level we will lose ourselves anyway, not by self-transcendence but by eternal self-destruction; we will fall into the world where everything is defined by the ego in its failed and despairing attempt to create and maintain itself—this being the state of hell, the 'darkness outside'). The second level of consciousness is where we lose our souls for His sake; this corresponds to the second level of consciousness posited by the Vedanta, the level of 'there is only Brahman', as well as to 'fana' (or 'annihilation') in Sufism. And the third level, the one where, because we have lost our souls for His sake, we now 'find' them, corresponds to 'I am Brahman', and to the Sufi 'baqa' or 'subsistence-in-God'. (The fourth Vedantic level, the level of 'All this, too, is Brahman', corresponds to everything Christians mean by APOCATASTASIS.) The same passage from self-defined subject through annihilation in the transcendent Divine Object to the unveiling of the Absolute Subject is also encapsulated in the hadith of the Prophet Muhammad, peace and blessings be upon him: 'Pray to God as if you saw Him, because even if you don't see Him, He sees you.'

And it is certainly true that an 'esoteric' ego is a much bigger and more savage beast than an exoteric one. 'To whom much has been given, much will be required.' All that is required of the exoteric believer—and it in itself is no picnic—is sincerity and humility; the esoteric jñani must submit to total annihilation and 'objectification'; if he fails in this he will rebel like Lucifer, and fall as just deep.

Sincerely,
Charles Upton

Dear Charles Upton

I am sure there must be plenty of theoretical arguments against what you have said, but more important than any of them is what G.K.Chesterton called 'the little dumb certainties of experience.' These are things which you appear to have exterminated so that you could contrive to put a world of phantoms between yourself and reality....

You speak so confidently about 'the Vedanta', although when we speak of it in relation to what I was talking about, this really only means the Vedanta as interpreted by Shankara. What we are talking about would not be possible, subject to the interpretations of Ramanuja or Madhva, I believe. Both Guénon and Schuon ignore that as well.

With what you say about all the levels to pass through, and the mysterious processes through which something or other (the ego?) must pass, it looks as though you have summed up the system of the truly real, but for one thing, namely, that YOU ARE THINKING IT, and that is something else again.

Yours Sincerely,
Robert Bolton

Dear Robert Bolton

You might as well say that the hill outside my window cannot be objectively real because I AM SEEING IT. It is true that I can never see it in exactly the same way that you could, but it is objectively real for all that; our differing perspectives converge upon it inexorably. If there is an objective metaphysical order as well as an objective sensual order (and Intellection would be no better than fantasy if there were not), then the fact that I am thinking about a metaphysical principle no more invalidates it than the fact that I am seeing the tree outside my window with my own eyes invalidates the objective existence of the tree.

Your point that I have unthinkingly identified 'the Vedanta' with the non-dualistic, Shankarian Vedanta alone is well taken. I will be more careful in the future. You ask if it is the EGO that goes through all those levels of consciousness. Good point! Language undoubtedly constrains us to speak as if the 'little me' were realizing God, as if it were capable of encompassing Him.

But as is made clear in the First Chapter of John, the little me cannot realize God, and insofar as that little me remains—which it always seems to do, at least for almost all of us, at least while we are still in this life—then it is a servant of God, in need of His grace and helpless without it. So the question is, are my ideas 'phantoms'? And if they are, are ALL ideas then phantoms? Or all metaphysical ideas? I suppose you mean to challenge me to ask myself whether all this metaphysical mumbo-jumbo is just a 'head-trip' on my part, whether it is simply a kind of information acquired by mental effort which lives nowhere but in my temporal memory which will perish with my mortal flesh. Undoubtedly some of it is. It is certainly possible to 'learn metaphysics' by rote—but it is also possible to speak out of realization, to express a metaphysical truth that is 'before one's eyes' in as concrete a way—in even more concrete a way—as the coffee cup on the desk. Assuming that I am sufficiently discerning and sufficiently honest about myself to say this, I would say that the metaphysical truths that I've written to you about are not a constant or stable realization for me, but come in flashes—by which I mean that I must draw partly on acquired knowledge, on the words of those wiser than I, to write about them. I am also willing to entertain the possibility that God, through you, is telling me to 'shut up', to do what I can for the rest of this life to make those realizations more stable, which is something that glib metaphysical expression can sometimes short-circuit—especially in my case.

Dear Charles Upton

Your e-mail struck me as a refutation of someone who had been so foolish as to set up the ego as a standard of value, which upset me because my conception of the real individual self is something incomparably greater than its ego. It is, I believe, a strong and original answer to the 'what am I?' question.

You challenge my use of the word 'impersonal' rather than 'transpersonal' in regard to the Absolute, but I can easily concede that, because it too is something known on the basis of the personal and not ontologically sepa-

rable from it. That is important for what we believe about God: whether God's unity embraces the personal, the transpersonal and the impersonal, or whether God-as-personal and God-as-transpersonal are ontologically different realities.

This is probably the essence of our disagreement. To begin with, you seem to come out against the latter alternative, referring to C.S.Lewis' remark about the pantheists who treat God as though He were just a book on a shelf. But are you not actually committed to a position for which the personal God who can require something of us is nevertheless the lesser of the two realities?

You say that all persons are 'personal faces of the same Transpersonal Mystery', when we should really say that they are all images of the archetypal humanity of Christ—and that humanity cannot be regarded as a mere mask, as the Monophysites believe. You say that Vedanta (according to Shankara?) does not divide reality between and Infinite God and an illusory cosmos, and that Maya does not mean 'illusion'; that is the opposite of everything I have been able glean on this subject, so the interpreters must be deeply divided. Do any creatures exist in their own right? That is very much an issue in modern Catholic theology, which may be influenced by the Vedanta in dividing everything between an absolutely self-existent God and an absolutely contingent creation. Being a Platonist, that means for me that in this case there could not possibly be any relation between them. All things are joined by means. Thus the highest members of creation share to a large extent in (created) self-existence, and only the lowest members are completely contingent. Hence the Great Chain of Being. Just to say that the soul is immortal is to say that it has a degree of self-existence. (Actually Aquinas affirms this).

On your fourth page, you say very emphatically that 'Who I really am can NEVER appear as an object of consciousness,' although it seems obvious to me that the whole of life brings a continually deepening awareness of that very thing. By holding this position, you undermine some other things you say: at the bottom of your second page you mention the four levels of consciousness through which Vedantists progress, but we cannot know what levels we are on if the 'I' cannot be an object of consciousness to itself. Similarly with what you say on your fourth page, where you say that we can only save our souls by losing them (into God's hands, presumably); we cannot know whether we are doing that or not, if the self is not an object of consciousness to itself. Perhaps you mean we just have to hope that we are doing so in the conduct of our lives. Losing one's soul in order to find it raises a logical problem, by the way: if we mean it literally, we are not really doing so, if we are hoping for anything—losing one's soul (literally speaking) must mean losing it in order to lose it. But in reality, the meaning of this expression is strictly of the moral order, not the ontological order.

The word Apocatastasis is used in a good many ways, but no Christians, apart, possibly Origen, have used it to mean that there will ultimately be nothing but God. That is Monism precisely.

Back to the 'two Gods' issue: the Transpersonal God who 'creates' a world by something as casual and contingent as causing a reflection, and the Personal God who creates the world on purpose and sees that it is good are either two antagonistic Gods, or this is all just a way of saying that, while God really is the Creator, He has many other activities which have nothing to do with it. If man can have a private life, why not God as well? I am a Christian Platonist, and that allows a certain kind of esotericism, which has very sound credentials. But the esotericism of Guénon and Schuon seems to me to owe too much to the 'tradition' founded by H.P.Blavatsky. We are clearly a long way apart, but I hope that this will help you to see what the issues between us are.

Yours sincerely,
Robert Bolton

Dear Robert Bolton

You ask me *whether God's unity embraces the personal, the transpersonal and the impersonal, or whether God-as-personal and God-as-transpersonal are ontologically different realities.* That is an extremely good question. I would say that It/He does embrace all these. Yet (paradoxically) we can still discern these ontological levels *within* that Unity (though to strictly identify them with the Persons of the Trinity is not warranted). I share what is perhaps your concern about some of Guénon's and Schuon's formulations of the ontological distinction between Saguna Brahman and Nirguna Brahman, or the Personal God as Pure Being and the Godhead as Beyond Being, which sometimes seem in danger of degrading the Personal God to some kind of independent, created demiurge. That is sometimes a problem with their language, though not (I trust) with their substance.

I certainly agree that we are not mere masks of the Transpersonal (despite the etymology of 'person' from the Latin for 'mask'; literally, something that is 'sounded through'); rather, we are individual *instances* of It: unique instances of the Human Archetype, which in itself 'opens up behind' into the Infinite. God Himself is not merely universal; He too is unique, is Uniqueness itself.

Do creatures exist in their own right? I would say that nothing created by God exists in its own right—at least in the same sense that God does—but for the fact that God confers that right upon it. In my own religion (Islam), the absolute sovereignty of God over creation—a sovereignty that is both willful and ontological—is emphasized to such a degree that some schools of thought seem to deny secondary causation. This, however, is not literally the case; no tradition that doesn't allow for secondary causation could have so advanced the human understanding of natural law. It's just that the First Cause is seen as absolutely superseding and dominating all secondary causes, though He chooses to allow them to operate, or actually wills them to operate. To say that God 'turns existence over' to secondary causes, however—which also implies that He turns it over to beings who (now) exist in their own right, and can thus create in their own right—opens the door to Deism. Sentient beings choose and create, natural laws operate, but always as created, witnessed, allowed, and ultimately willed by God. (When we will something, it is really God Who is willing it—NOT because we have no free will, but because, as we draw upon God's gift of His Own Being for our very existence, so we draw upon God's own power of willing for our actions and choices: if 'it is not I who live but Christ lives in me,' it is also not I who act, but God acts in me—action being an essential aspect of life. Yet He does not impose His Will upon us; this is the principle expressed by Ibn 'al-Arabi as 'the determined determines the Determiner.') So everything, on all levels of the Great Chain of Being, is absolutely contingent upon God; on pain of Deism we must assert

this. God's free gift of His own autonomy is the source of whatever autonomy we have, and such autonomy varies vastly in
degree, from that of the Seraph to that of the falling rock. Both
are equally contingent upon God in this present creative
moment. He wills both to be, and could at any moment will
them not to be. And both are totally free to be what they are as
he has made them. Yet the freedom and autonomy of the Seraph
immensely surpasses that of the rock.

[Bolton]: . . . *you say that we can only save our souls by losing them
(into God's hands, presumably); we cannot know whether we are doing
that or not, if the self is not an object of consciousness to itself. Perhaps
you mean we just have to hope that we are doing so in the conduct of our
lives. Losing one's soul in order to find it raises a logical problem, by the
way: if we mean it literally, we are not really doing so, if we are hoping for
anything—losing one's soul (literally speaking) must mean losing it in
order to lose it. But in reality, the meaning of this expression is strictly of
the moral order, not the ontological order.*

You are right in saying that if we *try* to lose our life *in order to*
find it, then we have defeated our own purpose—but then why
did Jesus say 'he who loses his life for My sake shall find it', if he
was not somehow *recommending* a kind of self-annihilation, of
which His crucifixion was the clearest and most complete
example? Jesus obviously knew that He would rise again, that
He possessed eternal life, but that didn't prevent Him from
going straight through the experience of 'My God, my God, why
hast Thou forsaken me?' Yes, we must lose ourselves completely
in God, as if we never knew that we were immortal, because we
cannot at the same time hold on to our desire for the continued
existence of our individual identity, and really let go of that
identity. We may plan to do this kind of letting go in the future,
in hopes of obtaining something infinitely better; the spiritual
life would be impossible without the theological virtue of Hope.
But when the moment of truth arrives, we have to (in Rama
Coomaraswamy's words) 'fish or cut bait'. In that moment, the
one we imagined as 'obtaining' something is no longer the old
'me', but rather one for whom nothing need be obtained
because, to him, all eternally *is*. The life we regain is His life, not

ours (and, in truth, it was always His). But since there is no continuity between my individuality and the Absolute, I really do have to die 'without hope', as hope is defined by my mortal thoughts and desires. Conversely, because there is nothing real that does not partake of the Absolute and is not supported by It, my human individual personhood is itself eternal—eternal as long as I have really died to it, died to its concupiscence, its pride, its temporality and its mortality.

Here you bring up one of the real paradoxes of mystical experience, or at least of the kind of mystical theology which says things like: 'I cannot know God, but God knows Himself in me.' You are absolutely right that, if the individual self cannot be an object of consciousness to itself, then there is no way we can experience ourselves as losing it to, or in, God. So the only way our life can be lost and regained in God, in full consciousness, is if the consciousness experiencing it is ultimately God's consciousness—not that of a strictly transcendent God, but of a God who is immanent, at this moment, in me. This immanent God is the Absolute Witness, the Atman: 'It is not I who live, but Christ lives in me.'

You say that *in reality, the meaning of this expression is strictly of the moral order, not the ontological order.* But *can* there be anything moral that is not, on another level, also ontological? If so, it wouldn't be *real.* (So much for the false *voluntaristic* exoterists who deny the Hierarchy of Being.) And can there be anything ontological that is not, on another level, also moral? If so, it wouldn't be *good.* (So much for the antinomian pseudo-esoterists.)

I say that the *exoteric* meaning of 'to lose one's life for Christ's sake' is *moral*, while its *esoteric* meaning is ontological. And the two are not ultimately separate (which, incidentally, is why both Elijah—symbol of the esoteric, ontological dimension—and Moses—symbol of the moral, the exoteric—appeared next to Christ in His Transfiguration). But who else says this? Probably not even Dionysius the Areopagite. Perhaps only Eckhart is explicit about it, when he says 'My truest "I" is God.' In my view, this is precisely the esoteric exegesis of 'he who loses his life for My sake shall find it.'

I define Atman as the 'I' Who knows Itself *essentially*—by *being* itself, not by becoming an object of consciousness to itself—which is not to say that It does not also (partially and imperfectly) become an object of consciousness to Itself, thereby manifesting the universe. In other words, we cannot say that before creation God was ignorant of His true nature, that he manifested the world as a kind of creative Self-exploration. *We* may learn more about ourselves in the act of creating something (though I believe that we often forget nearly as much at the same time, if not more), but God does not need to practice art therapy in order to better understand Himself; His Being *is* His Knowing.

You say: *The word Apocatastasis is used in a good many ways, but no Christians, apart, possibly Origen, have used it to mean that there will ultimately be nothing but God. That is Monism precisely.*

If by *apocatastasis* we mean that all things will be restored to their original form and stature as God created them, then this implies that human consciousness will also be so restored. I maintain that such restored consciousness sees all things *in* God. 'All this is Brahman' is not strictly Monism, since there is still an 'all this', and since, as I said, this fourth level does not negate the earlier three levels, but embraces them. This is what Schuon means by 'maya-in-divinis', and it is maya-in-divinis which negates strict 'literal' Monism. In the (Shankarian?) level two, 'there is only Brahman', all individual distinctions disappear; but here they are restored, as manifestations of God, not as veils hiding Him—as is the case with level one, 'Brahman is real, the universe is unreal.' Only a universe of veils need be negated in favor of God; a universe of theophanies need not be.

It seems to me that the experiential 'realization' of Monism is represented by those enraptured saints, much in evidence in India, to whom the particulars of the world, other people and themselves have disappeared, the ones immersed in *nirvikalpa samadhi* and therefore totally unable to deal with practical affairs. Traditions which recognize the existence and validity of such ecstatics usually speak of them as inferior to those sages who, while they may have passed through an ecstatic stage, have now 'returned' to the conditional, manifest world, seeing it all as a theophany, but not for all that ignorant of or unable to deal with the particulars of other people and of changing situations—even more able than most to deal with such things, some would argue, since they no longer view them through the obscuring veils of subjectivity. So Monism, though false as a description of the essential nature of things, does represent what the Sufis would call a *maqam*, a 'station', a proximate stage-of-realization, whereas the sage who sees all thing in God, in their total depth of particularity, but without this veiling to the slightest degree the Presence of the Absolute, sees things as they really are, and so is beyond all stations. This is what I believe is indicated by 'And all this is Brahman.'

Exactly how do you see Guénon and Schuon as part of an eso-teric tradition 'created' by *Blavatsky*? Doctrinally, which is what we are talking about here, they are poles apart. Of course Guénon investigated many occult organizations in his earlier life, and may have retained certain accidental habits of mind from those years, but that doesn't mean he shared any doctrinal common ground with the Theosophical Society, except by vir-tue of what the Society stole from the Vedanta (by 'stole' I mean 'appropriated for their own purposes, and in so doing took totally out of context'). Guénon wrote an entire book exposing the Theosophical Society as a pseudo-religion; are we to believe that he had no *doctrinal* reasons for doing so, that their disagree-ment was a mere turf-war among rival gurus? It is true that Blav-atsky on the one hand, and Guénon and Schuon on the other, spoke of a Primordial Tradition (in Hindu terms the *sanatanad-harma*), but this, in Blavatsky's rendition, is something that is des-tined to *replace* the revealed religions in the near future, while Guénon and Schuon maintained that it is manifest *in* the revealed religions, and is *only spiritually effective within the bounds of one of them*. This is the great divide between Blavatsky and Guénon. Guénon's teaching harks back to the days when God walked with man in the cool of the evening; Blavatsky's goes back only as far as the Tower of Babel. Guénon speaks (at his best) out of the Primal Word, Blavatsky only from the Confusion of Tongues. You can legitimately disagree with Guénon's position, but no well-informed person can confuse his teaching with Blavatsky's, except in accidentals.

You say: .... *the Transpersonal God who 'creates' a world by something as casual and contingent as causing a reflection, and the Personal God who creates the world on purpose and sees that it is good are either two antagonistic Gods, or this is all just a way of saying that, while God really is the Creator, He has many other activities which have nothing to do with it. If man can have a private life, why not God as well?*

Where you see 'casual and contingent', I see 'inevitable, effortless, and perfect'.

Do you think God has to *struggle* to create the universe? Wrestle painfully with His materials like some tormented artistic genius? Raise a sweat like a carpenter or bricklayer? As the Qur'an says, *He needs only to say to a thing 'Be!', and it is.* Once all is already created, in a moment of eternal time, *then* God's creative power extends further, toward bringing 'out' into existence what has been created, and finally toward working on, and with, what already exists. (Here we can see the operation, on three distinct levels, of three of God's Ninety-Nine Beautiful Names: the *Creator*, the *Producer*, and the *Fashioner*.) Perhaps on certain levels God is a Workman, but if we take the level where He (apparently) must struggle against the chaotic inertia of matter to build the cosmic harmony, then we deny His Omnipotence (besides starting to sound an awful lot like the Freemasons), which is instantaneous and, within the bounds of the ontological level upon which He is working, absolute.

Man truly would have a private life—if it were not for God. God is the only One whose private life is *absolutely* private. (I love your metaphor, by the way; it is worthy of C. S. Lewis.) Certainly He has many other 'activities', or modes of Being, than those defined by His role as Creator; and yet *all* His activities are subsumed under the definition of God as 'pure Act'. In God, all possibilities are actualized—by His nature, not by what we would think of as discreet, particular actions, in which what is first a mere potency is later made actual. The motion from potency to act happens in the already-projected *reflection* of God in the sea of cosmic *prima materia*, where the creative aspect of Him 'stands out' as Creator. Only when a Creator is confronted with a mass of possibility which has yet to be actualized can we speak of such a motion. But in the depths of the Godhead, all is actualized already.

The Creator, in essence, *is* the Godhead itself—yet, as you say, that Godhead also has a 'private life' beyond His creativity. To speak of this 'private life' is to speak of the Divine Essence per se—but truly there is no separation between this Essence and the Creator Who is Its manifestation—Its manifestation *to us* as creatures. The Creator is fully Godhead, and yet Godhead is not limited to Its creative function. God IS, Perfectly, in His Own nature—and since He Is (nothing else than) *that* He Is (to quote His Self-description at the Burning Bush), He is also Beyond Being. He is Beyond Being by the fact that He is neither this nor that; by the fact that, since He Is by His Own Essence alone, He is not one among the various things that possess Being; and by the fact that He is neither contingent upon some other Being, nor is He 'contingent upon Himself'. He has not created Himself and so need not maintain Himself. He need not BE. He is Beyond Being.

Sincerely,
Charles Upton

Dear Charles Upton

Thanks for your latest, but for this time will just say something about my original reply, which was cryptic and open to misunderstandings. Firstly, I had no intention of disparaging metaphysics as such—ideas are not just thoughts, but realities reflected in thought. In the 'world of phantoms' bit, I was projecting something of my own experience. In my younger days there was something which inclined me to Solipsism without my realizing it, and the Guénonian Vedanta blended with that, so I was a keen consumer of this Oriental mysticism. Solipsism gives one a world full of emptied beings, devoid of inner reality, inner worlds, or mystery. Those are the phantoms which one would be putting between oneself and reality. If you have no such problem with this doctrine, I would not know whether that was to the credit of your doctrine or of your psychic self-defenses. This is all of a piece with the 'little dumb certainties of experience'.

Monism/Non-Dualism, if taken seriously, has an effect of devaluing the reality of things we naturally take to be real, as though we could only make God look more real by making creation look less than real. That may be helpful for people who are inclined to make a God of the world, but do not include myself there. On the other hand, if the 'Illusion' doctrine is really just another way of underlining the difference between creation and the uncreated God, (as you seem to suggest), do we really need it?

Now the ego or the 'little me', as you call it: this perception of the self is wholly owing to sense-perception, which is deceptive in many ways, and most of all when it pretends to show us our own selves. We cannot base deep metaphysics, let alone initiatic knowledge, on sense perception and untrained common-sense perception at that. And yet, it seems to me that most of the impact of Vedantist mysticism depends on our taking this average man's sense perception of finite beings, passing into and out of existence like shadows, as though it were a revelation from God. But metaphysical knowledge must get behind these appearances, and the esotericism I have in mind does that. Conversely, if sense perception rules, it must define knowledge as such, and our metaphysical knowledge may well be phantasmal.

Here we get to the main focus of our differences: there are deeply different ways of defining the esoteric, and they depend in turn on how we define man himself. There are two diametrically-opposed ways of doing that, one of them of Indian origin, as adopted by Guénon and Schuon, and one of Egyptian origin. For the Indian doctrine, man is something which is really the same as God, but which gets all kinds of cosmic pollution stuck to it in the course of arriving in this world. So, then, we just have to scrape off the pollution, and there will then be nothing but God, just as it ought to be. This is practically the same as saying that man as such is not real at all.

The opposite of that is a conception for which this multi-levelled, microcosmic nature we have is not accretion, but is our very essence, created by God. Consequently, it would be self-contradictory to try to fully realize that essence by trying to be a pure spirit, like God. I could say more about the real or esoteric nature of the individual self, if there is any need for it. Possibly we have both gone so far on our different ways that changes of direction would do more harm than good, but we should both be made more aware of the reasons for what we believe.

Yours Sincerely,
Robert Bolton

Dear Robert Bolton

You say: *In the 'world of phantoms' bit, I was projecting something of my own experience. In my younger days there was something which inclined me to Solipsism without my realizing it, and the Guénonian Vedanta blended with that, so I was a keen consumer of this Oriental mysticism. Solipsism gives one a world full of emptied beings, devoid of inner reality, inner worlds, or mystery. Those are the phantoms which one would be putting between oneself and reality. If you have no such problem with this doctrine, I would not know whether that was to the credit of your doctrine or of your psychic self-defences.*

It seems to me that solipsism—most often the *unconscious* solipsism we call *narcissism*—is a very common way of misunderstanding the Vedanta, and esoterism in general. (I remember how Art Kleps, one of Timothy Leary's colleagues at Milbrook, Leary's 'psychedelic manor house' in upstate New York, seemed in his writings to have a very good understanding of the Vedanta and the Tantra—until he reduced it all to solipsism: he thought that the universe was just one big Art Kleps. This is one thing that can happen when you experience the voidness of all phenomena apparently due to some pill that *you*, and nobody else, have just popped.)

You say: *This is all of a piece with the 'little dumb certainties of experience.' Monism/Non-Dualism, if taken seriously, has an effect of devaluing the reality of things we naturally take to be real, as though we could only make God look more real by making creation look less than real.* [But this is *dualism*, isn't it?] *That may be helpful for people who are inclined to make a God of the world, but I do not include myself there. On the other hand, if the 'Illusion' doctrine is really just another way of underlining the difference between creation and the uncreated God, (as you seem to suggest), do we really need it?*

Again, those writers I follow—the Traditionalists, Ramana Maharshi, and others—do not define *maya* as 'illusion', but as the 'magical' self-manifestation of God. If we think that the universe is literally God, we are deluded by *maya* (*avidya-maya*), whereas if we see the forms of the universe as nothing other than manifestations of the very Godhead, *Who can never be defined in terms of that which manifests Him*, then we are witnessing *vidya-maya*. And this is the furthest thing from stumbling around in a world of contingent, illusory, dying phantoms: rather, it is a witnessing of the Real, as if face-to-Face, and of all forms as living, breathing instances of Life Itself. It is poles apart from an alienated vision of things; it is a vision of all things restored. In Blake's words, 'If the Doors of Perception were cleansed, all would appear to Man as it is, Infinite.' In the same vein, Olivier Clement, in *The Roots of Christian Mysticism*, quotes from Vladimir Maximov: '... it is as if I were seeing the forest for the first time. A fir tree was not only a fir tree, but also something else much greater. The dew on the grass was not just dew in general. Each drop existed on its own. I could have given a name to every puddle on the road.' In Buddhist terms, this is a vision of the union of 'suchness', *tathata*, and 'voidness', *shunyata*. This *shunyata* is not some horrible, dead emptiness—only *thought* thinks that. If is rather an 'emptiness of self-nature' in things, which means that things are not hidden from us, and from themselves, by some obscure sort of self-involvement, but rather *are* exactly as they *appear*, as they appear to the eyes of Truth Itself. They are not *mere* appearances; in all their uniqueness, they are the very appearance of the Real. This is their suchness, their *tathata*.

You say: *Now the ego or the 'little me', as you call it: this perception of the self is wholly owing to sense-perception, which is deceptive in many ways, and most of all when it pretends to show us our own selves. We cannot base deep metaphysics, let alone initiatic knowledge, on sense perception and untrained common-sense perception at that. And yet, it seems to me that most of the impact of Vedantist mysticism depends on our taking this average man's sense perception of finite beings, passing into and out of existence like shadows, as though it were a revelation from God. But metaphysical knowledge must get behind these appearances, and the esotericism I have in mind does that. Conversely, if sense perception rules, it must define knowledge as such, and our metaphysical knowledge may well be phantasmal.*

I'm sorry, but I just can't see how the Advaita Vedanta could be based on naive realism taken to its metaphysical extreme; that would be *materialism*. It is true that Hinduism and Buddhism—

Buddhism in particular—tend to emphasize the passing imper-
manence of things, but they do so *as a spiritual method*: If we witness our phenomenal selves as impermanent, for that precise
reason we will NOT identify with sense experience or seek to
hold on to old self-concepts, which are the chief among those
lifeless phantoms you mention. When the phenomenal is
allowed to pass, when it is clearly witnessed as passing, then the
Noumenon shines clear.

You say: *Here we get to the main focus of our differences: there are
deeply different ways of defining the esoteric, and they depend in turn on
how we define man himself. There are two diametrically-opposed ways of
doing that, one of them of Indian origin, as adopted by Guénon and
Schuon, and one of Egyptian origin. For the Indian doctrine, man is something which is really the same as God, but which gets all kinds of cosmic
pollution stuck to it in the course of arriving in this world. So, then, we just
have to scrape off the pollution, and there will then be nothing but God, just
as it ought to be. This is practically the same as saying that man as such is
not real at all.*

Here is where we run into the whole idea of reincarnation,
which Guénon, rightly or wrongly, denied was ever taught by
the legitimate Vedanta, and which may seem to deny the eternal
immortality (not the indefinite temporal extension) of the individual soul. Here is where we must ask the question: *what incarnates?*, which is another way of asking: *who am I?* Guénon
emphasizes the doctrine of the Advaita Vedanta that 'the Self (or
Brahman) is the one and only Transmigrant.' Only He passes
(apparently—apparently—) from form to form. On a less absolute level, however, we could say that what 'reincarnates' is my
physical and psychic *materia*, which, like my clothes and books,
may pass to new owners upon my death. My form, however, is
never repeated. And what is never repeated is, thereby, immortal from the standpoint of God's consciousness in the eternal
present. Exactly how Hinduism expresses this kind of immortality on a doctrinal level I am not entirely clear on—like most
westerners I jumped directly to the Vedanta; fool that I was, I
didn't want to busy myself with concepts that seemed *no better
than Christian*—but it *is* clear that the sages of the past, who can

intercede for the living, are thought of as immortal on some level; Ramakrishna himself had visions of those immortal exalted sages who are higher even than the gods. Likewise the dying Ramana Maharshi, when asked by his disciples where he would 'go' after death, replied: 'Go? Where could I possibly go?'

You say: *The opposite of that is a conception for which this multi-levelled, microcosmic nature we have is not accretion, but is our very essence, created by God. Consequently, it would be self-contradictory to try to fully realize that essence by trying to be a pure spirit, like God. I could say more about the real or esoteric nature of the individual self, if there is any need for it. Possibly we have both gone so far on our different ways that changes of direction would do more harm than good, but we should both be made more aware of the reasons for what we believe.*

The individual, form-bound self can never become pure Spirit (it is Luciferian to believe it can)—except as witnessed by pure Spirit, Who, while remaining totally aware of all form- and time-bound particulars, in essence (and paradoxically) sees nothing but Itself. The created, form-bound self remains a servant of its Creator. What happens, however—if it happens—is that the Absolute Witness is unveiled, after which point it is not I who witness myself, but He who witnesses me. I am objectified before the face of God, Who Alone knows me perfectly, precisely as I am. Before this *metanoia*, my contingent self is 'me' and God is 'He'. After it, my contingent self—is 'he', while God—is 'I' (Again, in Eckhart's words: 'My truest "I" is God.') In this state our individuality remains; it is not merged or blotted out in the Formless. Rather, it is *witnessed* by the Formless, which (paradoxical as it may seem) by Its witnessing both reveals my form-bound, witnessed self to be totally contingent upon the Formless, and at the same time sees it as being, in all its synthetic complexity, transparent to It. By becoming fully objective to the Absolute Witness, that self becomes most fully itself, and at the same time is revealed as fully 'void of (contingent) self-nature'. Its uniqueness is known as a unique instance of the Absolute Uniqueness of God. Thus what once appeared as various layers of cosmic accretion and pollution, obscuring the face of Truth—which is exactly what we seem to be (and, *effectively* though not

essentially, what we *are*) when perceptually limited by our passions and egotism—is now revealed as a manifestation of all it
once seemed to hide—a manifestation in which nothing is hidden. When veiled we are veils—when unveiled, unveilings—of
the Truth.

What you have to say is profoundly true to the dark, alienated
ways in which abstract thought can construe spiritual truth, and
to what so many westerners have actually found who have
turned to Buddhism and the Advaita Vedanta in flight from
Christ (in the case of Christians) or Yah-weh (in the case of Jews).
But Christ, Yah-weh are there too, if they only knew. And
undoubtedly many Hindus have the same alienated relationship
to their own tradition, as so many westerners clearly do to Christianity and Judaism, and so many Muslims to Islam. The Advaita
Vedanta too is old, in cultural terms. But the Truth Itself, though
it *is* old, never *becomes* old. The Ancient of Days is also, and simultaneously, the Ever-Young. Ibn al-'Arabi met him at the Kaaba.

Sincerely,
Charles Upton

Dear Robert Bolton

Theology is dogmatic; its field is *belief.* (I define 'theology' here
not in the Eastern Orthodox sense of 'the fruit of the spiritual
life in knowledge of God', but in the more western, Roman
Catholic sense of 'the official pronouncements of the Church
councils and the writings based upon them,' which—with the
'best opinion of the community of scholars' taking the place of
the church councils—is roughly equivalent to the Islamic
kalam.) Metaphysics is, on the other hand, is operative; its field is
intellective knowledge. This does not mean that metaphysical discourse can never employ theological dogma, or that dogmatic
truth can never be a catalyst for intellection. Yet intrinsic to
dogmatic theology is a kind of defensive function; is formulations, though they are the truest things that can possibly be said
(in dogmatic terms) about specific divine realities, are more for
the purpose of defending against error than directly expressing

Truth. Orthodox dogma as defense against theological error is something that can be established and finished with; it is the firm basis of all understanding, including the metaphysical; it is like a rock.

Metaphysical discourse, on the other hand, can never be finished with, never be established; it is like the wind; *it bloweth where it listeth*. Metaphysical discourse must walk a fine line between a total dedication to Truth and a realization that all metaphysical statements are 'operative' in a sense, ultimately justified only by the effect they produce on our consciousness. A statement that is 'true' but does not *convey* Truth in no way serves the goal of metaphysics—which is to prepare the heart not to resist, through attachment to error, the direct intuition of Divine realities. This is why the Buddha considered such questions as whether or not human consciousness survives bodily death, or whether the Tathagata does or does not exist after attaining Nirvana, as 'tending not to edification'. To the Buddhists, doctrine is an aspect of *upaya* or operative method. If it breaks our attachment to erroneous ideas, it is true; if not, then not. On the other hand, the clear, objective and orthodox expression of Divine Truth, even though it is incapable of totally encompassing the Truth it expresses, is one of the most powerful of *upayas*, which is why the Buddhists, even while repudiating metaphysics (as the Theravadins and the Zen practitioners do) also produced metaphysical systems of great subtlety and complexity (as, for example, that of the Vajrayana school, or the Madhyamika). To idolize metaphysical formulations (and theological ones as well) as if they were absolutely true, rather than 'quasi-absolutely' necessary within a given context, is to miss the Truth by the positive road, which is literalism; to take such formulations as mere utilitarian tools for the manipulation of consciousness is to miss it by the negative road, which is nihilism. Apophatic mysticism is the remedy for the first disease, cataphatic mysticism for the second.

Metaphysics is suggestive, not dogmatic; it must speak from a multitude of perspectives, since by nature it is attempting to express, or give intuitions of, something which cannot and should not be made explicit—a Reality beyond expression, beyond words. Metaphysical truth can and should be expressed in terms of logic, insofar as is possible—and such expression is far more possible than is generally believed—but it cannot be trapped in logic. Theology is the necessary protecting and limiting vessel of metaphysics; the metaphysical multiplicity of perspectives is no excuse for heterodoxy; metaphysics is not simply a license to say anything you want to; it is not subjective or impressionistic; some formulations are simply wrong. Its specific function is to protect the providential and necessary formulations of theology from petrifying into idols, and also to point beyond them, directly to the Object of which they are the established signposts in this world.

I am a Muslim whose practice is Sufism. I was raised a Catholic in an essentially pre-Vatican II Church, given that the radical changes wrought by that Council took years to percolate down to the parish level. I am a lifelong student of 'comparative' religion and metaphysics; beyond that, I also have a background in the 'counter-culture' of the 6os and 7os.

I hold to Frithjof Schuon's doctrine of the Transcendent Unity of Religions, according to which each true, God-given religion is 'relatively absolute', and contains all that is necessary for salvation (speaking in moral and personal terms) and liberation (speaking in intellective and transpersonal terms). I am firmly opposed to syncretism, and recognize that a comparative study of religion and metaphysics from the standpoint of the Transcendent Unity of Religions is always in danger of becoming syncretistic, consciously or otherwise. Furthermore, I do not attempt to escape from the potential contradictions of the doctrine of the Transcendent Unity of Religions by saying, 'doctrinally I am a universalist; I am an exclusivist only in practice', because the orthodox doctrines of a given religion are in fact an intrinsic aspect of the practice of that religion. What we believe determines what we do, and is in itself a form of intellective action.

On the other hand, there is a way in which I am a doctrinal universalist, in that I accept as true the doctrines of all the revealed religions; where they contradict each other—which is both inevitable and providential on the plane of form—I have trained myself to refer them back to their Transcendent Source rather than setting them at war with each other within my consciousness, trusting that in the next world (mysteriously present in the heart of this one) God—in the words of the Qur'an—*will enlighten us as to wherein we differ.*

Why, then, do I not simply limit myself to the doctrines of my own religion, Islam, seeing that I have just admitted that each religion contains all that is necessary for salvation and liberation? First, I must simply declare that, earlier in my life, my *cultural* horizon (in religious terms) inevitably expanded beyond the *cultural* limits of Christianity (and Islam), and I have seen no sufficient reason to shrink it again. We all, in fact, do live in a 'pluralistic' world where many spiritual traditions meet, clash, intermingle, misrepresent each other, and (God willing) enlighten each other, and it would do no good for me to pretend otherwise. Furthermore, I have recognized in my 'heart' the central truths of the major world religions—which does not mean that I have identified with them or simply adopted them as beliefs, nor does it mean that I have reached perfection, sainthood, or even psychological stability; to recognize the truth of something is not necessarily to fully actualize that truth. But once you do recognize the truth of something, it dominates you; you become the slave of it. You may be a sullen, rebellious slave or a faithful and willing one, you may attempt to live by what you know or try your best to deny what you know, but slave of that recognized truth you must remain, in this world and the next.

What, then, is the use of 'comparative religion', if not simply to serve academic secularism, or syncretism? I would say that each religion has its unique points of incomparable excellence, where certain facets of the Truth—the real Truth, the Truth no religion has ever captured or ever will (no one seriously thinks of God as a 'Hindu' or a 'Christian'), the Truth which brought forth the religions in their first beginnings and encompasses them even now—are revealed in a more perfect and effective way than in any other tradition. The essence of Islam is submission to the Absolute; of Christianity, that God as Love, Love even unto death, and beyond death; of Hinduism, that all exists only as witnessed by That which can never be witnessed; of Judaism, that to be bound by the God-given Law is freedom, and election, and wisdom; of Taoism that by 'doing without doing' we are at one with the Way of all things; of Buddhism, that no thing is a thing-in-itself but rather an instance of original Awakeness, and to realize this is the end of suffering. The perfection of each of these religions can be found in all the others—but a particular perfection can perhaps only be fully recognized in terms of the light that another religion can throw upon it. Islam takes 'not my will but Thine be done' and places it at the Center; the Hindu doctrine of the Atman illuminates both the Christian 'it is not I who live, but Christ lives in me' and the Muslim 'he who knows himself knows his Lord.' The Buddhist doctrine of *annata* deepens the meaning of the Muslim 'I (God) am the Rich (in Being) and ye are the poor' (God here being equivalent to the Buddhist 'Void eternally generative', eternally *generous*). The Sufi doctrine that all things are submitted to God *intrinsically* deepens the meaning of what it means to be 'at one with the Tao', while Taoism itself is the highest and deepest rendition of 'the evil of the day is sufficient thereunto' and 'consider the lilies of the field, they neither sow nor do they spin, but never was Solomon in all his glory arrayed like one of these.' Each revelation has a different Center, but each Center also contains all other Centers; that's what a Center is. The Center is everywhere, the circumference nowhere.

But what does all this mean for me in *practice?* How is saying 'within the Center are all Centers' not syncretism? If the Truth which gives rise to all the religions is One, still their forms differ, and these differences are irreducible—so how do I deal with them? How can I accept them all without blending and confusing them? How can I avoid confusing them without being partial to one of them ('partial' as opposed to *whole*)? In fear and trembling, I hazard to express it like this: Though I know that Christ is God and the Son of God, I do not *use* this knowledge; as a Muslim, the knowledge I *use* is that God is One and Sufficient unto Himself, that He neither begets nor is begotten, and that there is nothing to which He may be compared. And if this appears to you as the most shocking sort of arrogance and nihilism, if not outright stupidity, then so be it; it is simply the truth as I see it, the truth (or delusion) by which I will sink or swim. I have both a right and a duty to use this knowledge, since it is this unique rendition of Truth, this specific Name of God, by which God has called me, at least during this phase of my life, to worship Him; it has nothing to do with my subjective 'preferences'. (What I *preferred* would have led me straight to Hell, if God had not intervened.) In the words of the Muslim daily prayer, I pray that God lead me on a straight path, and that I not find myself among those who have earned His anger, or those who have gone astray.

Sincerely,
Charles Upton

Dear Charles Upton,

In arguing for the Guénonian idea of the Vedanta, you are arguing for a form of mysticism which requires that the world should result exclusively from emanation and not creation, and so this discussion will have to be mainly concerned with the question as to whether there is a real creation or not. Your latest reply begins by agreeing that God is really one, and so comprises everything personal and transpersonal, so here is at least one orthodox belief about God on which we are agreed.

This could well create problems for the doctrine of Illusion, however, and the Non-Dualist idea of 'identification', as Guénon called it. The 'two Gods' idea, Illusion, and 'identification' appear to stand or fall together. If God cannot be divided into one half which is above Maya and one which comes under it, because the determinate and the indeterminate aspects of God are equally real, God must be outside this conceptual framework. Why then may not everything else be?

You have said before that 'Maya' does not really mean 'illusion', but in that case, you are parting ways with Schuon. He affirms that it does so in Chapter 5 of Gnosis: Divine Wisdom, and he says that God as Creator is 'determined by Maya' in his *Survey of Metaphysics and Esotericism, p.* 55, and for him 'Maya', 'illusion', and 'collective dream' all mean the same thing. He relates this to 'All is Atma', which implies that everything but the Absolute must be illusion.

In any case, you would be right to discard the doctrine of Illusion, because it is really just a monument to bad logic. Its premise is that the world cannot be as real as God is, but illusion does not follow from that. There are so many relevant examples, e.g., as homes go, mine is not much compared with Buckingham Palace, but that does not mean it is not a home, or that it is anything less than a home; practical arithmetic is not as advanced a subject as algebra, but it is in no way less mathematical. In fact, God gives everything the exact amount of being and reality which is appropriate for it.

Where you say that 'we are individual instances of It,' you appear to be excluding the idea that we are individually created, and this is a case where an implicit emanationism is evident. You state that 'nothing created by God exists in its own right.' Not in the sense of self-created, of course, and most modern people are completely blind to this. But you add that if God 'turns existence over' to secondary causes, that 'opens the door to Deism'. But it is a principle in Scholastic tradition that God never acts directly in the world when He can act through some subordinate agent. Since you reject the idea of God creating the world laboriously, you should accept this idea of causal delegation.

When St. Paul said 'it is not I who live, but Christ lives in me,' he was not saying that he was Christ, or that he was incapable of sin or error, but the use you make of this text does imply both that and that we could say the same of ourselves. But if we deny our independent being or substances, we may well have to so speak of ourselves. Nevertheless, the idea of created beings with some degree of absoluteness is to be found in Aquinas: 'But to be simply necessary is not incompatible with the notion of created being....'

Again: 'The more distant a thing is from that which is a being by virtue of itself, namely, God, the nearer it is to non-being; so that the closer a thing is to God, the further it is from non-being.... Thus some created beings have being necessarily.' *Summa Contra Gentiles, Vol.II,* Chapter 30, (5) & (6). This text excludes the idea that while God has absolute self-existence, everything else has only absolute contingency (if that could be more than nothing). There are many degrees of real and necessary being in creation, and that is why it manifests its Creator. God does not make rubbish, and Moslems and Christians should be able to agree about that.

Later on, you speak of 'a kind of self-annihilation, of which His Crucifixion was the clearest and most complete example,' but He only submitted to that and did not do it to Himself. When He rose again He was still Himself, with the same recognizable human personality, the Wounds, and the same relationship to the disciples. If that was annihilation, we all get annihilated every time we go to sleep at night.

The idea that 'we must lose ourselves completely in God,' can be understood in a moral sense without an ontological one as well. One can be 'selfless' in a moral sense, and still have a real self physically, and one can be 'of one mind' with one's associates when all have real minds of their own. If you insist on the idea of 'losing oneself' in the ontological sense, you are ignoring the distinction between the corrupt and selfish self, and the virtuous self, and treating them both as the same kind of evil to be eradicated equally, as though God made a mistake in creating, and as though nature was evil in itself. This results from following a doctrine which does not distinguish between the Creation and the Fall, and which therefore has something unmistakably Manichaean about it. In any case, what are affirming here could only mean physical death.

This does not mean that I am unaware of the need for theoretical truth to be 'realized', but I have a different idea as to what that means, but one which is still traditional. It is more like a slow 'assimilative conversion' which calls for more patience than modern people can find. This belongs naturally with a Platonic intellectual basis, and the point of departure for that is the immortal soul, and to pretend that that did not exist would be as pointless as pretending that God did not exist. On this basis, 'losing oneself' cannot have the same meaning, because the 'losing' would have to be something going on in the self or soul that is supposed to be getting lost. Your doctrine applies to the self's ego, not to the self or soul.

I have worked out some consequences of this in 'Dualism and the Philosophy of the Soul', which appeared in Sacred Web 4. These were opposed to the Guénonian-Vedantist idea of the individual self, because the latter is tied to the common sense idea of self-as-ego, and that the world containing the ego is made up of things just as one perceives them. These perceptions are assumed to be shared by everyone else, and to be the causes of their experiences, just as much as of one's own, when they are in any case only the final effects of the objective world, and not causes at all. Each person is thus conceived solely as an object contained in a common world which is really only one's own. If each soul in reality contains its own representation of the world which contains its ego, the self-annihilation idea must mean something quite different. It need not be crudely untrue, of course, but just more limited and more psychological.

But if our psycho-corporeal being is thus, must it not still be mortal and perishable, even if the soul itself is immortal? Possibly even that is only an appearance which deceives the senses. Esoterically, the person who appears to perish is really only a minute part of the real person which extends as a continuum through countless states over different times. That would not be literally open to annihilation on any natural level. I gave some explanation of that in 'Life, Death and Resurrection', in Sacred Web 7. This is the reason why I said that the Non-Dualist doctrine depends on an idea of reality based only on sense-perception. Besides, it is so largely expressed in terms of temporal processes that we should be warned not to take it too literally for that reason alone.

The position you are defending would be sound enough if Eckhart's saying 'My truest "I" is God,' were literally true just as you quote it, but in reality the literal sense could only amount to one of two absurd alternatives, either: (1) (assuming that we are real beings) that God was divided up into little bits (by whom, or by what?) for the private use of creatures, in which case God could not be separate from creation; or: (2) that God is not really divided, and the world and ourselves are just illusions, and there is no creation. Phantoms only.

But what Eckhart was really referring to here was not precisely God, even though it is in many ways Divine, but what is known as the 'eye of the soul', or the 'Divine spark', or the 'synteresis', of the soul. This is necessary to complete the human microcosm, and its nature is the next reality to God, that of the Uncreated Heaven. I give an explanation of this in Chapter 8 of my next book which is in preparation at the moment, but if that had to be adapted for popular consumption, one would end up saying that the highest part of the soul was 'God', just as Eckhart did. When we judge statements, we must also consider those to whom they were addressed before we take them literally.

Then there was my remark about H.P. Blavatsky and the Traditionalists. I do not see why the idea that Guénon's break with the Theosophists as

'a turf war among rival gurus' should be so obviously wrong. The Spirit-
ist Fallacy is mostly an attack on the shortcomings of other Theosophists
and not an attack on Theosophism itself, if I can read French. That is why it
is so tedious—four hundred pages of hacking away at other people's inadequacies, and no attempt to engage with the fundamental issues of Theosophism. It all amounts to proof of a lot of very soured personal relationships
and disillusionments, I would say.

It is very likely that, when Guénon ceased to practice his Catholicism, he
compensated by intensifying the inner mindset behind it, the passionate
interest in orthodoxy and fear of losing it. He may thus have seen heretics
everywhere and ignored the possibility that he was one himself. Similarly,
many Protestants who lost their faith used to compensate by working
harder at Christian morality. This is a well-known religious phenomenon.
At any rate, Guénon ended up claiming to speak for orthodoxy from a position ABOVE orthodoxy, rather as the Theosophists were doing in their own
way. We have no way of knowing that this was not just the latest version of
the Serpent's 'Ye shall be as gods,' so I do not see how you can feel so certain that he was right.

Guénon was in some respects a typical intellectual of his time, influenced
by the way in which metaphysical thought in Europe had already moved in
the direction of monistic Pantheism under the influence of 19th Century
German idealism. Blavatsky just had to connect that with the Shankaran
Vedantic tradition. That meant that European intellectuals would assume
from then on that the most monistic interpretations of Indian thought must
be the most authoritative.

Where creation is concerned, and you ask 'Do you think that God has to
struggle to create the universe?' that would be clearly anthropomorphic, but
something of that problem remains even when we say the He creates it with
the greatest of ease, because that is human as well. Some degree of anthropomorphism is in fact necessary in view of man's 'Theomorphism', as
Schuon would call it. As for effort, I suppose Omnipotence must be able to
create challenges for itself, as the state of the modern world seems to confirm. One objectionable aspect of the 'reflection' analogy of creation is that
it excludes the idea of there being any intention to create any specific beings.
The personal relation between God and man would therefore have no basis,
and the saving of the soul would be purely and simply a matter of human
activities, as Guénon evidently thought it was.

Another objection to it is that it explicitly confuses the creation of the world with the procession of Christ from God the Father: 'He reflects (N.B.!) the glory of God and bears the very stamp of his nature, upholding the universe by the word of his power.' (Heb. 1:3, Catholic R.S.V.) If this can be said of the processions of the Trinity, it can NOT be said of the natural world—except on the emanationist assumption that there is no essential difference between God and creation.

The same view of God and creation appears where you say 'Here we must ask the question: what incarnates?' The one coherent consequence of the reincarnation idea is that there are no personal identities, because all apparent identities are just so many ephemeral disguises for God. So to affirm reincarnation is to affirm the doctrine of Illusion and deny the reality of creation, although both our faiths affirm the reality of creation.

Similarly, where you say that 'The individual form-bound self can never become pure Spirit,' you are denying theoretically something which Non-Dualism affirms in the concrete with its belief in an 'identification' of the self with God, who is necessarily a pure spirit. But regardless of any supposed reincarnation, man's not being a pure spirit is not an accident, since he is specially created as an epitome of all levels of being. Each person is therefore a world, so that the real objective world consists of the sum total of all these worlds, and of which one's own world is just a tiny part. Such is the true Macrocosm, known only to God. (The smallness of the ego in relation to its world reflects this on the empirical level).

If you can say that you are aiming for a state where 'my contingent self is "he," while God is "I",' this would mean that God is either unwilling or unable to confer real being on anything. One may, of course, say that God wants to confer on us something much more important, namely, Himself, but even Omnipotence cannot confer itself on nothing. There must be a common measure between the recipient and the Received, and that is why created personal identities must be real, and why they must have appropriate degrees of infinity in them, as instanced in their comprehension of worlds.

The Vedantist (and Moslem?) perspective is so much centred on man-becoming-God that there is no room for the complementary Christian doctrine of God-becoming-man. This issue must make a very serious difference to what it is that one is supposed to 'become!'

Yours sincerely,
Robert Bolton

Dear Robert Bolton

You say: *In arguing for the Guénonian idea of the Vedanta, you are arguing for a form of mysticism which requires that the world should result exclusively from emanation and not creation, and so this discussion will have to be mainly concerned with the question as to whether there is a real creation or not. Your latest reply begins by agreeing that God is really one, and so comprises everything personal and transpersonal, so here is at least one orthodox belief about God on which we are agreed.*

This could well create problems for the doctrine of Illusion, however, and the Non-Dualist idea of 'identification', as Guénon called it. The 'two Gods' idea, Illusion, and 'identification' appear to stand or fall together. If God cannot be divided into one half which is above Maya and one which comes under it, because the determinate and the indeterminate aspects of God are equally real, God must be outside this conceptual framework. Why then may not everything else be?

God is outside all conceptual frameworks. In Maya, in manifestation, I meet Him as the personal God, person-to-Person. When 'I' have gone beyond manifestation by the realization of the Atman—when I am no longer the subject looking to God as the Object, but when He is the Subject, witnessing me and all things as object—then He is entirely transpersonal and without attributes. While I am alive in this world, and while I have not realized the station of *jivanmukhta*, there the subjective 'me' who sees God as 'Thou' must remain on some level. So the 'division' of God into 'the two Gods' has nothing to do with God in His Essence, but rather with the spiritual station of the one conceiving Him, or—in the case of the realized Atman—the one conceived by Him. The Sufi al-Hallaj, speaking out of an ecstatic state in which his sense of his own existence was annihilated, said: 'I am the Truth', a statement for which he was martyred. Since in that moment 'al-Hallaj' was annihilated, the Speaker was in fact God Himself. Likewise the Sufi Abu Yazid (Bayazid) said: 'There is nothing beneath this cloak but God.' The great Sufi shaykh Junaid, however, while crediting the truth of these statements on a certain level—a level the Sufi's call 'drunkenness'—emphasized that we do not go about among men naked, without our 'cloak' (the human psycho-physical personality); on the level of our contingent humanity, which is always there in some sense while we live, we remain created, contingent beings and servants of our Lord.

You have said before that 'Maya' does not really mean 'illusion', but in that case, you are parting ways with Schuon. He affirms that it does so in Chapter 5 of Gnosis: Divine Wisdom, and he says that God as Creator is 'determined by Maya' in his Survey of Metaphysics and Esotericism, p 55, and for him 'Maya', 'illusion', and 'collective dream' all mean the same thing. He relates this to 'All is Atma', which implies that everything but the Absolute must be illusion.

But are dreams entirely illusory? The scientist who finally solved the structure of the benzene ring, one of the bases of organic chemistry, did so on the basis of a dream. It was through a dream of the Pharaoh interpreted by Joseph that he was released from prison, made a high official in Egypt, and was thereby able to save not only his brothers and his father Jacob from famine, but the whole Judaic tradition. And another Joseph, through another dream, was able to save his son Jesus from the soldiers of King Herod, and flee with him and his mother Mary into Egypt. When Schuon says 'the world may be a dream, but it is not our dream,' he means that the world, in relation to me, is objectively real; it is not simply a solipsistic projection of my own self-enclosed consciousness. And I can't agree with you that Schuon says that all things are literally illusory without any qualification. In *Gnosis: Divine Wisdom*, Chapter 5, he says:

> That all is *Atma*—or *Maya*, according to the point of view— clearly does not authorize us to 'take a rope for a snake' as the Vedantists say; quite the contrary.... Certain theorists of the Vedanta, anxious to buttress their conviction of the exclusive reality of the Self, 'inner witness' of all thought, feel obliged to deny the reality of the object as if it were 'mind' that creates the objective world—the Scriptures teach the contrary.... [note 2] 'The appearances (external objects) are not caused by the mind, nor is mind the product of appearances.' [*Mandukya-Karika*].

Like you, I am not entirely comfortable with Schuon's statement that the Personal God is 'determined by *Maya*'; it seems to imply that, since he can know this, he encompasses God rather than being encompassed by Him. And while he does say in a footnote (*Gnosis: Divine Wisdom*, p. 58n1) that Maya also means 'universal unfolding', 'divine art' and 'cosmic magic', he also maintains that the meaning 'illusion' is 'incontestable'. Elsewhere, however, he defines Maya in a more positive way. For example, in the article 'Atma-Maya' (*Studies in Comparative Religion*, vol. 1, no. 3), he says 'we could define [Maya's] nature by a multitude of different terms such as Relativity, Contingency, Separativity, Objectivization, Distinctivization, Exteriorization... even the term "Revelation" could apply here in an altogether fundamental and general sense', and maintains that '*Maya* is not only an "illusion", as the Advaitins would have it, but also a necessary adjunct of the goodness inherent in the Absolutely Real.' I believe that such apparent inconsistency has to do with the suggestive rather than dogmatic quality of metaphysical as opposed to theological language.

But if we take Maya as 'illusion', then what does it mean, in terms of the spiritual life, to say that the Personal God is a function of Maya? First, we must be clear that the Personal God is infinitely above my personal self, that He is the very face of the Absolute as directed toward that self. Furthermore, the Absolute—as I have said above—cannot be strictly 'impersonal', otherwise the Personal God could not be It's first intelligible hypostasis. It might be permissible to say that the Absolute is 'Personhood without limitation.' For us, personhood entails limitation; if I am myself, then I can't be you. But God is Personhood Itself (in Schuon's words, 'the Absolute Subject of our contingent subjectivities'), a Personhood beyond the limitation of form; He is not 'this as opposed to that' or 'this person as opposed to that person.' How can we conceive of an Infinite Personhood not limited by form? Perhaps only by realizing that such a Personhood has already conceived of us, and that His conception of us is in fact our truest nature.

So I am perhaps departing from Schuon here, perhaps not. Is an inconceivable, Infinite Person what Schuon means by the Personal God? Or does he mean by that term our *experience* of that God, which is necessarily limited by our own form-bound personhood? Ramana Maharshi says that the Personal God, Ishvara, does indeed, in literal reality, create, maintain and rule the universe, but from the standpoint of the ultimate realization, the Supreme Identity, He is simply the last thought—which means that He is vastly realer than the universe of which I am a part; I certainly do not arrogate reality to myself rather than to Him when I call Him 'illusion'. It's simply that the God we contemplate as an Absolute Object, or as an Object which *stands* for the Absolute, is necessarily limited—*as our experience*, though not in His Essence—by our own limited personhood as contemplators. And once we realize this, we understand that any experience of God is what Ibn al-'Arabi calls 'the God created in belief'—a subjectively limited image of God which God infuses with His Own Reality, thus making it worthy of worship by virtue of the Reality for which it stands. In His Reality, God is beyond our experience, beyond all experience, and therefore beyond form; if He were not, He would in no way transcend us, any more than the cow in the pasture. We see that cow out there, we know where it is and what it is; not so with God. The problem here is that some misguided people keep trying, in contemplation, to *stare into the Formlessness of the Divine Essence, as if Formlessness could be intelligible.* In other words, they substitute *their idea* of Formlessness for their idea of the Personal God. One's idea of God as Person is alive with Divine Reality, but one's idea of the Formless Absolute is nothing but a nihilistic void; to try and contemplate Formlessness *as an object of one's own consciousness* is to stare into the abyss of the ego. That's why Sufis are prohibited from attempting to contemplate the Essence; it simply doesn't work, and it is filled with many dangerous pitfalls. How, then, is our subjective vision of the Personal God to be transcended? How can we know God as He really is, rather than simply as we are forced, by our own limitation, to imagine Him to be? Our limited selves can never encompass God: but we *can* experience ourselves as *being encompassed by Him.* As Abu Bakr said, 'To know that God cannot be known is to know God.' First we pray to God as if we saw Him; then we realize that we can't see Him; finally we know Him as the One who sees us, and that we are nothing, in essence, but the objects of His knowing. So to say that 'the Personal God is a function of Maya' is to understand that any possible experience of God as an object of our limited consciousness is conditioned by that very limitation—a limitation that can only be transcended when we give up our role as subjective knowers and allow ourselves objectively to be known; this is the unveiling of the Atman, the Absolute Witness. And only someone who is called to this degree of realization has the right to assert, and the need to understand, that 'the Personal God is a function of Maya.' To say 'God is an illusion' will be taken by most people to mean that God is 'only' an illusion, or that He is 'my' illusion, or the illusion of some collectivity of souls. Those to whom God is not quite real—and that includes most believers, however sincere—must first be impressed with the overwhelming REALITY of God. Only those who know God as Reality Itself can make fruitful use of the idea that God *as experienced* is illusion—the word 'illusion' here obviously denoting, as I believe it did for those rishis who first elaborated the doctrine of Maya, the manifestation of a Reality which so infinitely transcends our experience of It, no matter how glorious, tremendous, and loaded with meaning that experience may be, that *even this*, even the wondrous universe, is mere illusion in comparison with the Reality it manifests. In other words, apophatic mysticism does not negate cataphatic mysticism, but rather frees it from limitation. That's India. But when the doctrine of Maya is taken up by Europe, the Europe that produced Sartre and Kafka, it is in danger of being interpreted as positing 'a world so drab and lifeless that it isn't even real.' Both India and Shakespeare agree that 'All the world's a stage', but where India sees the play as God's *lila*, Shakespeare (perhaps speaking for a quality that began to enter the European psyche with the Renaissance) called it 'a tale told by an idiot, full of sound and fury, signifying nothing.' That's how important the imponderables of cultural context are in determining the effective meaning of words like *Maya* or *illusion*.

It might seem here that to explain Schuon's statement that the Personal God is a function of Maya *by virtue of our necessarily*

limited perception is to deny Schuon's contention that He is a function of Maya *intrinsically*, and furthermore to imply that He is
'our dream', which Schuon also denies (since if the world itself is
not our subjective projection, still less can we say this of God).
The answer to this objection is that the Maya that manifests the
Personal God also manifests us as limited personal beings who
must intrinsically see Him as a formally-limited and therefore
intelligible Person. God indeed creates us, but does so by
'breathing the Mercy of existence' (according to the Sufi conception of it) upon the eternal possibilities (the *ayan al-thabita* or
'permanent archetypes') that already exist within His own
nature, which Schuon terms 'Maya-in-divinis'. In this sense we
are woven into the same web; only the Supreme Identity, the
Infinite Person, transcends the limitations of that web. And if
the power that makes Him what He is (the *shakti* that is one in
Essence with the Personal God, while being the full and complete consort only to the Transpersonal God, to Nirguna Brahman), and the power that constrains us to see Him the way we
do, are one and the same, then the Personal God cannot be
defined as an illusion of our own subjectivity.

You say: *In any case, you would be right to discard the doctrine of
Illusion, because it is really just a monument to bad logic. Its premise is
that the world cannot be as real as God is, but illusion does not follow
from that. There are so many relevant examples, e.g., as homes go, mine is
not much compared with Buckingham Palace, but that does not mean it is
not a home, or that it is anything less than a home; practical arithmetic is
not as advanced a subject as algebra, but it is in no way less mathematical. In fact, God gives everything the exact amount of being and reality
which is appropriate for it.*

To say this is to imply that God is only relatively greater than
you or me—but the fact is, He is infinitely greater. He is just not
the greatest thing that happens to exist, like the tallest mountain
or the richest man on earth. Whatever is, He transcends it; this
is what Muslims mean by *Allahu Akbar*: not just that God is great,
but that whatever may actually or conceivably exist, God is
greater than that. According to the Sufis, to say 'there is no god
but God' is not simply to say 'there is only one God,' but that

'there is only one Reality'. This would be monism, except for the fact that (paradoxically, it would seem) God is also the Creator, and the universe He creates certainly possesses a relative reality—real insofar as it manifests Him, unreal insofar as it seems to be able to exist apart from Him.

Where you say that 'we are individual instances of It,' you appear to be excluding the idea that we are individually created, and this is a case where an implicit emanationism is evident. You state that 'nothing created by God exists in its own right.' Not in the sense of self-created, of course, and most modern people are completely blind to this.

But you add that if God 'turns existence over' to secondary causes, that 'opens the door to Deism'. But it is a principle in Scholastic tradition that God never acts directly in the world when He can act through some subordinate agent. Since you reject the idea of God creating the world laboriously, you should accept this idea of causal delegation.

Certainly God acts through agents, both spiritual (angels) and corporeal (natural laws). Yet their actions are still His. They do not act apart from Him; He acts *through* them; *they themselves are His acts.* When God establishes a natural law, such as the law of gravity, in that creative moment He wills all the separate instances of the operation of that law; a sparrow does not fall without His knowledge and will. He does not simply say, 'Hmm, let's set up this Law of Gravity thing and see what happens.' He *knows* what happens, what will happen. If saying that God 'turns existence over' to secondary causes implies that these independent agents can *surprise* him, like our children often surprise us, then the door is open not only to deism but to process theology, which is the denial of both God's omniscience and His omnipotence; through the unexpected actions of his creatures, God too is growing, developing etc. As a Muslim, I cannot accept this.

Secondary causes exist as a 'subset' of the First Cause; whatever is caused by God indirectly in terms of secondary causes, is caused by Him directly, and eternally, in terms of the First Cause. Sufism does not deny secondary causes, but rather deliberately adopts the standpoint of the First Cause, this being an essential aspect of Muslim piety. For example, in the face of what is sometimes called 'a stroke of fate', an instance of great good fortune or great misfortune, Sufi practice, and Muslim practice in general, is not to consider secondary causes, but rather to refer such events directly to the First Cause, to thank God for them, or resign ourselves to God's Will in the face of them, not (ultimately) to enquire into the secondary causes through which that Will may have acted; in the presence of any event conceived of as the direct action of God, such considerations are spiritually irrelevant. We can make perfect submission, perfect *islam* to God's present act, but we cannot do so to some concatenation of secondary causes. Therefore, in the face of inexorable events, we don't say 'I wonder how that happened, how that might be repeated, how that could have been avoided' (though the practical wisdom that on its own level must ask such questions is never denied); we say: 'it is God's will; glory be to God.'

When St. Paul said 'it is not I who live, but Christ lives in me,' he was not saying that he was Christ, or that he was incapable of sin or error, but the use you make of this text does imply both that and that we could say the same of ourselves. But if we deny our independent being or substances, we may well have to so speak of ourselves. Nevertheless, the idea of created beings with some degree of absoluteness is to be found in Aquinas: 'But to be simply necessary is not incompatible with the notion of created being.... Again: 'The more distant a thing is from that which is a being by virtue of itself, namely, God, the nearer it is to non-being; so that the closer a thing is to God, the further it is from non-being.... Thus some created beings have being necessarily.' SCG, Vol. II, Chapter 30, (5) and (6).

This text excludes the idea that while God has absolute self-existence, everything else has only absolute contingency (if that could be more than nothing). There are many degrees of real and necessary being in creation, and that is why it manifests its Creator. God does not make rubbish, and Moslems and Christians should be able to agree about that.

Certainly St. Paul was not saying that he was Christ or was perfect (remember the 'thorn in my flesh'); he expressly says 'not I'. (But what, then, is theosis? Can you enlighten me? And what are we to make of Christ's command 'be ye perfect, even as your Heavenly Father is perfect'?)

And certainly Islam does not deny the Great Chain of Being; the 'Five Presences' of Ibn al-'Arabi are one of the most sophisticated renditions of this chain. As for 'degrees of necessity' in created beings, that wholly depends upon what we mean by 'necessity'. The fact is that God is necessary to me, but I am not necessary to Him; I cannot exist without Him, but He can exist without me. On the other hand, if all possible beings and events were not already actualized within God—this being Schuon's doctrine of 'Maya-in-divinis'—He would be less than God. Because God is Necessary Being, therefore, and because His Being is Infinite, I too am necessary in a sense; the rock on the hillside is necessary in the same sense. Nonetheless I do not create Him, but He creates me.

This level of reality, however, is not the one that deals with *degrees* of necessity. In terms of the Great Chain of Being, we could say that each higher ontological level is necessary in relation to all that is below it, and contingent in relation to all that is above it. But in terms of the Sufi path, it is not my level of relative necessity that is emphasized, but rather my nothingness in the face of God, Who is Absolute Necessity; in the face of Absolute Necessity, I am quasi-absolute contingency; I am effectively nothing. Yet through the empty doorway that very nothingness, God makes a gift to my nothingness of His own Being—thus my contingency *is* my necessity, my self-annihilation *is* my eternal life. This is the Sufi doctrine of *fana* and *baqa*, annihilation and subsistence. I know that there are logical problems with expressing it this way, but there are also logical problems with 'it is not I who live, but Christ lives in me.' (If *I do not live*, then there is no *me* for Christ to *live in*, etc., etc.) So unless we want to call St. Paul simply muddle-headed, we will have to recognize that such statements come from a type of consciousness where logic no longer strictly applies.

Later on, you speak of 'a kind of self-annihilation, of which His Crucifixion was the clearest and most complete example,' but He only submitted to that and did not do it to Himself. When He rose again He was still Himself, with the same recognizable human personality, the Wounds, and the same relationship to the disciples. If that was annihilation, we all get annihilated every time we go to sleep at night.

I entirely agree that Christ did not commit suicide; but did

He not submit *actively* to His Father's will that He die on the cross? Could He not have avoided the cup he elected to drink by calling on, at the very least, twelve legions of angels? He *willed* his own death on the cross by conforming His will to His Father's will; if we deny this, if we say that the Father's will simply over-powered Christ's will, we risk the heresy of monothelitism.

The idea that 'we must lose ourselves completely in God,' can be under-stood in a moral sense without an ontological one as well. One can be 'self-less' in a moral sense, and still have a real self physically, and one can be 'of one mind' with one's associates when all have real minds of their own. If you insist on the idea of 'losing oneself' in the ontological sense, you are ignoring the distinction between the corrupt and selfish self, and the virtu-ous self, and treating them both as the same kind of evil to be eradicated equally, as though God made a mistake in creating, and as though nature was evil in itself. This results from following a doctrine which does not dis-tinguish between the Creation and the Fall, and which therefore has some-thing unmistakably Manichaean about it. In any case, what are affirming here could only mean physical death.

Here you raise a very important point: that we must not con-fuse the Creation and the Fall by asserting or implying that existence itself is evil. True, the great Sufi woman saint Rabi'a said, in answer to a young man who told her that he had never committed a sin, 'Alas, thine existence is a sin with which no other sin may be compared.' Taken *literally*, this is heresy, imply-ing that God can create evil and thus in some sense *is* evil. It is not meant as a literal statement, however, but as an operative one: To the degree that the young man, filled with spiritual pride, saw himself as self-created, or saw his holiness as a per-sonal achievement, he was close to what Jesus called 'the sin against the Holy Spirit'. (How similar this encounter is to that of Jesus with the 'rich young man'.)

It is true that we are created in the image and likeness of God. It is also true that we are fallen. And no amount of expansion or improvement or purification of our fallen selves can reverse the effects of the Fall. We must die to those selves, which means, effectively, that we must die to all we believe ourselves to be. In Christian terms, we are redeemed not simply by accepting some good moral advice, or some effective spiritual practice, or a certain amount of grace from God through Christ; no: we must *die with* Christ and *rise with* Him. And Christ didn't simply die to His corrupt self while retaining His virtuous self; He *had no* corrupt self, but was all virtue. Entire self-annihilation is the only way back from the fallen self to the self as God created it; if it weren't, we would be like William James's 'once-born' men, accepting the good of creation and their own human nature without any radical encounter with the Fall, and thus without any (apparent) need for sacrificial redemption. If 'losing ourselves' were the end of the story, then we would indeed be Manichaeans or Gnostics or nihilists of one kind or another to say that the self should be lost; but there is also 'finding ourselves', in and for God. There is not only crucifixion and death, there is also resurrection. But simply to know that resurrection is the end of the story does not allow us to *skip* the crucifixion, but rather it calls us to go through with it.

This does not mean that I am unaware of the need for theoretical truth to be 'realized', but I have a different idea as to what that means, but one which is still traditional. It is more like a slow 'assimilative conversion' which calls for more patience than modern people can find. This belongs naturally with a Platonic intellectual basis, and the point of departure for that is the immortal soul, and to pretend that that did not exist would be as pointless as pretending that God did not exist. On this basis, 'losing oneself' cannot have the same meaning, because the 'losing' would have to be something going on in the self or soul that is supposed to be getting lost. Your doctrine applies to the self's ego, not to the self or soul.

Well... yes. But to the soul identified with its own ego, the *experience* of self-annihilation is the experience of death to all it knows itself to be. Using physical death as a metaphor for ego-death, we may firmly believe that our soul is immortal and that death is not the end, and still be terrified at the approach of death, as if it were the end of everything, thus proving that what we *really* believe is not always the same as what we *believe* we believe. And 'losing the self' does not go on within the self any more than drowning in the ocean goes on inside my stomach. How could it? The whale really does swallow Jonah, for all that he vomits him up again later on. Jonah fled from God's command into his own ego, and was therefore forced by God to experience the non-existence *intrinsic* to that ego.

I have worked out some consequences of this in 'Dualism and the Philosophy of the Soul', which appeared in Sacred Web 4. These were opposed to the Guénonian-Vedantist idea of the individual self, because the latter is tied to the common sense idea of self-as-ego, and that the world containing the ego is made up of things just as one perceives them. These perceptions are assumed to be shared by everyone else, and to be the causes of their experiences, just as much as of one's own, when they are in any case only the final effects of the objective world, and not causes at all. Each person is thus conceived solely as an object contained in a common world which is really only one's own. If each soul in reality contains its own representation of the world which contains its ego, the self-annihilation idea must mean something quite different. It need not be crudely untrue, of course, but just more limited and more psychological.

Once again, I do not believe that the Hindus deny individual immortality, though they see Liberation as superseding it. They pray to the *pitris*; they posit many after-death hells and paradises; they do not simply see the individual *jiva* as a function of matter and sense-perception.

But if our psycho-corporeal being is thus, must it not still be mortal and perishable, even if the soul itself is immortal? Possibly even that is only an appearance which deceives the senses. Esoterically, the person who appears to perish is really only a minute part of the real person which extends as a continuum through countless states over different times. That would not be literally open to annihilation on any natural level. I gave some explanation of that in 'Life, Death and Resurrection', in Sacred Web 7. This is the reason why I said that the Non-Dualist doctrine depends on an idea of reality based only on sense-perception. Besides, it is so largely expressed in terms of temporal processes that we should be warned not to take it too literally for that reason alone.

Very interesting, and undoubtedly true. It reminds me of Blake's 'the body is the portion of the soul perceived by the five senses.'

The position you are defending would be sound enough if Eckhart's saying 'My truest "I" is God,' were literally true just as you quote it, but in reality the literal sense could only amount to one of two absurd alternatives, either: (1) (assuming that we are real beings) that God was divided up into little bits (by whom, or by what?) for the private use of creatures, in which case God could not be separate from creation; or: (2) that God is not really divided, and the world and ourselves are just illusions, and there is no creation. Phantoms only.

But what Eckhart was really referring to here was not precisely God, even though it is in many ways Divine, but what is known as the 'eye of the soul', or the 'Divine spark', or the 'synteresis', of the soul. This is necessary to complete the human microcosm, and its nature is the next reality to God, that of the Uncreated Heaven. I give an explanation of this in Chapter 8 of my next book which is in preparation at the moment, but if that had to be adapted for popular consumption, one would end up saying that the highest part of the soul was 'God', just as Eckhart did. When we judge statements, we must also consider those to whom they were addressed before we take them literally.

In one sense the synteresis is on a lower ontological level than the Divine; in another sense, it is continuous with It. An open doorway is not the world outside the house, but is continuous with that world and is in no way a second object apart from that world. The synteresis is equivalent to the Vedantic *buddhi*, which in turn is analogous to the scholastic Intellectus. The Vedanta makes a distinction between buddhi and Atman, the former being like a first reflection of the latter. But I would say that buddhi is in a sense continuous with Atman, as the space of the doorway is continuous with the world. Christian theology is careful to guard against the identification of buddhi or synteresis, conceived of as the center of the human microcosm, with God, lest the necessary border between limited, form-bound man and the Formless Absolute be breached, leading to a Luciferian inflation of the human level, an attempt to 'be like God'. (How this caution squares with 'God became man so that man might become God' is something I hope you can explain to me.) Yet when Meister Eckhart calls Intellectus 'uncreated and uncreatable', is he not, in one sense, identifying it with God? What else can be called 'uncreated'? Synteresis considered as a level of the human microcosm is in no way God; considered as the indwelling Essence of all things, uncreated and uncreatable, it *is* God, or at least His Presence, which *is* Him but does not *exhaust* Him. The doorway as a framework of brick and wood, six feet high and three feet wide, is in no way the universe. As an open space, it is the actual *presence* of the universe, distinguishable from it in one sense, indistinguishable in another. It is both measurable and immeasurable. So I would say that the aspect of synteresis which is indistinguishable from God is what the Vedanta means by Atman. Atman is distinguishable from Brahman in one sense, indistinguishable in another; it is Brahman in the context of *jiva*. Jiva is host to Atman but does not imprison It. The human soul is host to the Imago Dei, Who in Christian terms is none other than Christ, a Christ who is both 'true God' and 'one with the Father', but it cannot be identified with Him. In the words of the *hadith qudsi*, 'Heaven and earth cannot contain me, but the heart of my believing slave can contain Me.'

Then there was my remark about H.P. Blavatsky and the Traditionalists. I do not see why the idea that Guénon's break with the Theosophists as 'a turf war among rival gurus' should be so obviously wrong. The Fallacy of Spiritism is mostly an attack on the shortcomings of other Theosophists and not an attack on Theosophism itself, if I can read French. That is why it is so tedious—four hundred pages of hacking away at other people's inadequacies, and no attempt to engage with the fundamental issues of Theosophism. It all amounts to proof of a lot of very soured personal relationships and disillusionments, I would say.

Yes, perhaps; but all possible personal disillusionment and bad blood apart, their doctrines really do differ.

It is very likely that, when Guénon ceased to practice his Catholicism, he compensated by intensifying the inner mindset behind it, the passionate interest in orthodoxy and fear of losing it. He may thus have seen heretics everywhere and ignored the possibility that he was one himself. Similarly, many Protestants who lost their faith used to compensate by working harder at Christian morality. This is a well-known religious phenomenon. At any rate, Guénon ended up claiming to speak for orthodoxy from a position ABOVE orthodoxy, rather as the Theosophists were doing in their own way. We have no way of knowing that this was not just the latest version of the Serpent's 'Ye shall be as gods', so I do not see how you can feel so certain that he was right.

Guénon was in some respects a typical intellectual of his time, influenced by the way in which metaphysical thought in Europe had already moved in the direction of monistic Pantheism under the influence of 19th Century German idealism. Blavatsky just had to connect that with the Shankaran Vedantic tradition. That meant that European intellectuals would assume from then on that the most monistic interpretations of Indian thought must be the most authoritative.

Where creation is concerned, and you ask 'Do you think that God has to struggle to create the universe?' that would be clearly anthropomorphic, but something of that problem remains even when we say the He creates it with the greatest of ease, because that is human as well. Some degree of anthropomorphism is in fact necessary in view of man's 'Theomorphism', as Schuon would call it. As for effort, I suppose Omnipotence must be able to create challenges for itself, as the state of the modern world seems to confirm.

Perhaps this is what Islamic theology, that of al-Ghazali, for example, is getting at by hierarchicalizing the Names of God: God as *Creator* need only say to something 'Be!' and it is; God as *Producer* 'leads forth' something that already *is* on another level; and God as *Fashioner* is metaphorically like a craftsman working on a material that is in some sense separate from him.

One objectionable aspect of the 'reflection' analogy of creation is that it excludes the idea of there being any intention to create any specific beings. The personal relation between God and man would therefore have no basis, and the saving of the soul would be purely and simply a matter of human activities, as Guénon evidently thought it was.

God does not choose between alternatives, He wills what is. He does not wonder what might have happened if He had taken a different road. As Pure Act, He is without what we see as 'activity', which is in reality the realization of possibilities, the passage from potency to act. We, who are involved with possible being, must experience God as choosing the actions He will perform, as saying 'no' to some petitions and 'yes' to others. But in His own nature, His Action is indistinguishable from His Being. The problem here is that too often we think we can 'access' God in His own nature without passing through God as one who acts, chooses, says 'yes' and 'no' *in relation to us*, to His creation; this is one of the meanings of 'none come to the Father but through Me.' There are so many pseudo-esoterists who think they can realize the Godhead while not even granting God the power (in relation to them, and to the universe) of acting on His own; to them He is like some kind of passive, unconscious natural resource they can 'tap' for spiritual power. To engage in petitionary prayer, for example, seems like some kind of shameful, backward superstition to people like this; what they really believe in is their own individual intelligence and spiritual self-determination, not the Mind and the Will of God.

However, I don't see Guénon as in this category. If he had believed that self-willed action could attain salvation and liberation, he would not have written of the necessity of membership in a traditional religion, and (on the esoteric level, and in strictly Islamic terms) initiation into a Sufi *tariqa* with a chain-of-transmission stretching back to the Prophet. It is these resources that are the source of the grace or *baraka* without which spiritual advancement is not possible, and anyone who posits the necessity of grace can't say at the same time that one can reach God by simple human action. That's the difference between Blavatsky and Guénon: Guénon was a *traditionalist*, Blavatsky a *promethean occultist*.

Another objection to it is that it explicitly confuses the creation of the world with the procession of Christ from God the Father: 'He reflects (N.B.!) the glory of God and bears the very stamp of his nature, upholding the universe by the word of his power.' (Heb. 1:3, Catholic R.S.V.) If this can be said of the processions of the Trinity, it can NOT be said of the natural world—except on the emanationist assumption that there is no essential difference between God and creation.

In Catholic high school in the mid-60s I was taught that God holds the universe in existence by His attention to it; if He were to withdraw that attention, it would sink into nothing. Is this no longer taught? Or is it in fact heterodox? Is it possible for God to forget about the universe and let it go off on its own? I'm sure you'll agree that it is not. It seems to me that if God sees the created world *sub specie aeternitatis*, then in a certain sense HE IS CREATING IT NOW, 'upholding the universe by the word of

His power.' Following the metaphor of the Universe as God's reflection, He has the power to turn toward it and see His Image within it; He also has the power to turn away from it, at which point his cosmic image (though not his Divine Image, his Son) immediately vanishes. So He certainly wills the appearance of His Image in, and as, the cosmos, but He need not labor to construct it.

And emanationism is not to be strictly identified with monism; the universe-as-emanated steps down the Great Chain of Being into manifestation; each successive emanation is on a lower ontological level than the previous one. Yet the Sufis maintain that the Essence (not the form) of what is emanated into those descending orders of existence is still the One Reality. In the words of the Holy Qur'an, *God made the universe with nothing but the Truth—if you only knew.*

The same view of God and creation appears where you say 'Here we must ask the question: what incarnates?' The one coherent consequence of the reincarnation idea is that there are no personal identities, because all apparent identities are just so many ephemeral disguises for God. So to affirm reincarnation is to affirm the doctrine of Illusion and deny the reality of creation, although both our faiths affirm the reality of creation.

Yes—though Guénon denied that essential Hinduism teaches reincarnation (the accuracy of this statement can certainly be disputed!) Once again, it is clear that the physical material of our bodies can pass into other bodies, and that our ideas can enter other minds; but our eternal form, being unique, cannot reincarnate; in Paradise, after the resurrection, this form is clothed in a new and incorruptible matter, like the 'glorified body' of Jesus after His resurrection.

Similarly, where you say that 'The individual form-bound self can never become pure Spirit,' you are denying theoretically something which Non-Dualism affirms in the concrete with its belief in an 'identification' of the self with God, who is necessarily a pure spirit. But regardless of any supposed reincarnation, man's not being a pure spirit is not an accident, since he is specially created as an epitome of all levels of being. Each person is therefore a world, so that the real objective world consists of the sum total of all these worlds, and of which one's own world is just a tiny part. Such is the true Macrocosm, known only to God. (*The smallness of the ego in relation to its world reflects this on the empirical level*).

Perhaps I am seeing the Vedanta in a distorted way through the eyes of Sufi metaphysics, but I don't believe that it actually teaches that *jiva* is to be identified with Atman. It is host to Atman intrinsically, and becomes transparent to Atman in the case of the *jivanmukhta*, but that which is limited to a particular form cannot be strictly identified with the Formless Absolute, though it may certainly be termed a manifestation of that Absolute according to the power of Maya. What *is* taught is that Atman is to be identified with Brahman. In Christian terms, though the soul is not to be identified with Christ, the Christ who 'lives in me' *is* 'of One Substance with the Father'. But perhaps I am wrong here about what the Vedanta actually teaches; could you cite any Hindu scripture which asserts that jiva, not Atman, is identical with Brahman? I would certainly agree that each person is a world, and that the real objective world consists of a sum total of these worlds; I would simply say that God creates or emanates (depending upon the point of view we adopt) these worlds, which, as instances of possible being contingent upon His Necessary Being, are, in *one* sense, 'illusory'.

If you can say that you are aiming for a state where 'my contingent self is "he", while God is "I"', this would mean that God is either unwilling or unable to confer real being on anything. One may, of course, say that God wants to confer on us something much more important, namely, Himself, but even Omnipotence cannot confer itself on nothing. There must be a common measure between the recipient and the Received, and that is why created personal identities must be real, and why they must have appropriate degrees of infinity in them, as instanced in their comprehension of worlds.

To say that man as contingent self can become God is a heresy in Islamic terms. When al-Hallaj said 'I am the Truth', that's what his fellow Muslims thought he meant, and so they executed him—and rightly so! If that's what he was really saying, then he deserved to die (if anybody actually deserves to die for the crime of heresy, which can certainly be disputed, though on a much different level).

This thing of 'objectification before the Absolute Witness' is hard to describe. It has nothing to do with the contingent human soul becoming God, or 'like God' (to quote the serpent), any more than it does with the literal illusory nothingness of that soul. Have you ever felt that you were known by someone, at least for a moment, more completely than you knew yourself? That's the beginning of it. The feeling that you are seen, that you are an object of the consciousness of another more fully than you are an object of your own consciousness. This can momentarily happen in situations of romantic love, and it's part of what's *supposed* to happen in psychoanalysis. When the Prophet Muhammad, peace and blessings be upon him, was asked by someone what was the perfection of the Islamic religion, he answered: 'Pray to God as if you saw Him, because even if you don't see Him, He sees you.' Ordinarily we take it as axiomatic that we see God, or might see Him, or that we can at least partly conceive of Him. In the same way, we think we know who we are, who someone else is, and what the world is, because *we are the knowers*; everything else is an object of *our* knowledge. But what if *being known* were much more powerful and real to us than *knowing*? What if 'He sees you' were a present, living experience so real that it undeniably and firmly subordinated our own feeble attempts to know ourselves, and know the God Who is seeing us? We would walk down the street as if followed by a great searchlight shining upon us from behind. Whenever we turned around to look at that light we would not be able to see it, since it would move as we moved, remaining always behind us. What we would see would be the world before us as illuminated by that light—that, and our own shadow cast upon that world. (Here we perhaps begin to see a deeper dimension to the allegory of Plato's cave.) But what would that *light itself* see, if it were also an eye? It would see the world, one a tiny part of which would be us. While we are identified with ourselves, even if we experience ourselves as being seen by God, we still exist, in a sense, as the shadow of our own self-involved subjectivity cast upon, and obscuring, the world around us. But from the point-of-view of God's light, which is the Atman, we are transparent. From our own perspective we cast a shadow; this is our ego. But from the perspective of the divine light, nothing casts a shadow; all things are seen as they are. (This was also a prayer of the Prophet: 'O God, show me things as they really are.') This is what it really means to 'die to oneself'. What dies is our self-reflexive subjectivity, our narcissism. When we as subjective selves attempt to see or understand ourselves, we necessarily see only 'in part'; the narcissist—and everyone who has not died to himself is a narcissist to one degree or another—sees only the mirror-reflection of himself, or of part of himself; he sees 'as in a glass, darkly'. But when the function of witnessing passes to (or is revealed as always having been intrinsic to) the Absolute Witness, then we no longer exist as partial images of ourselves; we are fully revealed as the wondrous microcosms we are, containing and epitomizing all levels of being. (The microcosmic and hierarchical nature of the human form is fully accepted and expressed in Hinduism through the doctrine of the psychophysical nervous system with its *chakras*, which are more than just elements in a kind of subtle or mythic human anatomy, but also represent actual levels of being, and function as doorways to the actual worlds occupying these levels.) This individual microcosm is both revealed in all its hierarchical splendor, *and* seen as no more than a tiny shrimp in the ocean of universal manifestation. (Plato's cave is, precisely, egoistic subjectivity; the great light outside is the Absolute Witness, the Atman. Are we now beginning to pick up the trail of a synthesis between Egypt and India?) These microcosmic selves are certainly not God, because God is the Witness, the 'Father.... Whom none has seen at any time.' Nor is the witnessed self in any way an 'illusion' in the sense of a 'phantom'. While I exist as no more than a set of subjective impressions of myself, maybe then I am a phantom, or the closest thing a human being can come to phantomhood. But when I exist as seen by God, when God's objective view of me (my immortal soul), not my own subjective view of myself (my ego), is revealed as the real me, then all phantoms are dispelled.

You say: *The Vedantist (and Moslem?) perspective is so much centered on man-becoming-God that there is no room for the complementary*

Christian doctrine of God-becoming-man. This issue must make a very serious difference to what it is that one is supposed to 'become!'

In Hinduism the avatars of Vishnu are, precisely, 'God become man'. And certainly Christianity has the doctrine of *theosis*, of man-become-God, which would at least seem to be analogous to the Vedantic *moksha*. What, in your opinion, is the difference, if any, between *moksha* and *theosis*?

Actually Islam is much more reticent to speak about 'man become God' than either Christianity or Hinduism. Only the Sufis do so, and usually not in so many words; al-Hallaj and Bayazid are exceptions even in Sufi terms. And Islam, at least exoteric Islam, expressly denies that 'God becomes man', this being *hulul*, the heresy of 'incarnationism'. On the other hand, the Sufi exegesis of the Quranic verse 'God created the universe only with the Truth—if you only knew' is to say that God, intrinsically, is the Essence (not the form) of all things, seeing that 'the Truth' (*al-Haqq*) is a Name of God's Essence. And the doctrine of *al-Insan al-Kamil*, the Perfect Man, who is primordial as well as ultimate, and is thus the prototype of the universe *in divinis*, embraces the idea of a Divine Humanity. All orthodox Muslims, Sufi or otherwise, though they call Jesus not simply a prophet but 'a Spirit from God', deny that Jesus is 'the Son of God', seeing that (in the words of the Qur'an), *God neither begets nor is He begotten*. Muslims never tire of emphasizing the Uniqueness and Incomparability of God, and abhor the idea that He could somehow beget a 'second God', which idea they often (erroneously) attribute to the Christians. However, to say 'Jesus Christ is the Son of God' is to speak *metaphorically*. If he were *literally* the Son of God, it would mean that God has or had a wife, that he had sexual intercourse with her, got her pregnant, and nine months later she gave birth to Jesus Christ. This would make it necessary to say that Mary was God's earthly wife, and that Mary in a higher sense, or some other divine hypostasis, is his wife in Eternity, the one who eternally gives birth to the Son begotten by the Father. But since the Christians do not in fact say this, the Muslim denial that Jesus was God's Son in this literal sense is not necessarily an absolute denial of the Christian mystery. Both

Christians and Muslims agree that God is One, and that God does not beget a second or third God. And both should certainly be able to agree that the term 'Son of God' must mean something other than literal human sonship. I am not denying that Jesus Christ is the Son of God; I am saying that the term 'Son of God' is *metaphorical*. As with most theological language, the term 'Son' is used—in my opinion—to emphasize that Christ is 'begotten, *not made*' and 'of one substance with the Father'; and in so doing to defend against heresies like Arianism; it is not meant to assert that God produced him through intercourse with a divine partner (to ascribe partners to God is the Muslim sin of *shirk*, equivalent in some ways to the Christian 'sin against the Holy Spirit'), or that Jesus Christ is literally God's 'Son' in the sense that, like the son of an earthly father, he might some day become a father himself and beget further sons (i.e., further gods). Christians relate to God through the mystery of His Trinity, while not denying His Unity; Muslims relate to Him directly through the Mystery of His Unity, while allowing room for certain trinitarian expressions—at least among the Sufis, as for example the chapter on 'Salih' in Ibn al'Arabi's *Fusus al-Hikam*.

Sincerely,
Charles Upton

Dear Charles Upton

Just to add something to the idea of 'illusion': Schuon, in Gnosis: Divine Wisdom admits that the usual arguments for it are not very good, and then goes on to offer one of his own for it, which is one of the worst he ever used. This is the idea that we are all in a collective dream because our archetype Universal Man, is 'dreaming', and thus making us all dream with him. Considering that we are speaking of an eternal Form, this is like saying that the Square on the Hypotenuse is feeling sick. He must have known that archetypal and physical realities could not be conflated like that.

This idea involves a denial of our own faculties which may as well be a denial of our sanity. It is utterly irrational to think that we shall wake up and get real when some big mystical revelation comes. Why should that not be a dream as well? Besides, if we can only dream that we have a desire for reality, we probably do not have one. As Schuon himself says somewhere, a being who is inherently absurd cannot have the possibility of ceasing to be so.

Although we may try the alternative of avoiding the word 'illusion' by speaking of 'the magical self-manifestation of God', and being able to 'see the forms of the universe as nothing other than manifestations of God,' we not much further on. At best, this is a way of speaking about heightened forms of experience, such as poets have, but even so we do not have a right to take such things as encounters with God, because God has no sensory or sensuous attributes. In such experiences, what is really happening is that the mind is struck, not by God as such, but by the eternal Forms of the things observed. They are experienced so intensely that the individual properties of their instantiations are forgotten. But God is not the world of Forms, even though He is ultimately the cause of the relationship of our faculties to the formal causes of the world. In any case, the difference between 'illusory' and 'magical' is mainly semantics, and there is no great mileage in it.

My attitude, you suggest, is owing to 'the dark, alienated ways in which abstract thought can construe spiritual truth.' Nevertheless, I construe a lot of other kinds of spiritual truth with the same mind, but with very different results. If conceptual thought indicates something dark and alienated in connection with Non-Dualism, it would be more natural to assume that that is the fault of Non-Dualism; but if we must allow at least the possibility that it is owing to this kind of thought, one still cannot assert that it is so without proof. If one does, it could only mean that Non-Dualism is being put beyond the range of discussion, that is, set up as a dogma.

Another reason as to why logical analysis might find something negative about Non-Dualism, from a Christian point of view, is that it requires one to discard a doctrinal truth, i.e., that the world is a real thing, created by a personal God Who transcends it. Having done that, it would not be surprising if the new outlook appeared dismal. It is far easier to be an ex-Christian than to acquire another and commensurate truth. 'The way down is easy,' but the way back up may not be recognizable.

There is a problem with Indian thought here, which is not just mine. Europeans are spiritually Egyptians—the Judaeo-Christian tradition, Platonism, Pythagoreanism, and Hermeticism are all of Egyptian origin. The fact that God acted in history through Moses and in Christ clearly did not alter its essential nature. Therefore, trying to graft Indian spirituality onto people with that heritage is really just a way of increasing the amount of confusion in the world.

Yours sincerely,
Robert Bolton

Dear Robert Bolton

You say: *Just to add something to the idea of 'illusion': Schuon, in* Gnosis: Divine Wisdom *admits that the usual arguments for it are not very good, and then goes on to offer one of his own for it, which is one of the worst he ever used. This is the idea that we are all in a collective dream because our archetype Universal Man, is 'dreaming', and thus making us all dream with him. Considering that we are speaking of an eternal Form, this is like saying that the Square on the Hypotenuse is feeling sick. He must have known that archetypal and physical realities could not be conflated like that.*

Schuon here is using the word 'dream' *metaphorically.* He doesn't mean that God dozed off after dinner and helplessly dreamed the universe. By the word 'dream' he means 'a world created out of one's own substance, with no immediate perceptual relation to any outside reality.' When we, as limited and imperfect beings, have dreams, these dreams are highly subjective (though as I pointed out above, they can also contain certain objective elements) and are thus less real than the waking world around us. But when God dreams, His dreams are as real as can be; not as real as Himself, of course, but in no way less real than some other more objective world, because—given that God creates *ex nihilo*—there is no such other world.

This idea involves a denial of our own faculties which may as well be a denial of our sanity. It is utterly irrational to think that we shall wake up and get real when some big mystical revelation comes. Why should that not be a dream as well? Besides, if we can only dream that we have a desire for reality, we probably do not have one. As Schuon himself says somewhere, a being who is inherently absurd cannot have the possibility of ceasing to be so.

Well, it seems here that you deny the universal testimony of the mystics. When St. Paul was hit by God's light on the road to

Damascus, though his fleshly eyes were blinded by it, the eye of his Heart was opened. Given that he passed from being a persecutor of Christians to the premier Christian missionary and proto-theologian, I don't know how you can call that anything other than a great awakening. And as for dreams, I can only quote the words of William Butler Yeats to the effect that 'In dreams begin responsibilities.'

And where does Schuon say that human beings are 'inherently absurd'?

You say: *Although we may try the alternative of avoiding the word 'illusion' by speaking of 'the magical self-manifestation of God,' and being able to 'see the forms of the universe as nothing other than manifestations of God,' we not much further on. At best, this is a way of speaking about heightened forms of experience, such as poets have, but even so we do not have a right to take such things as encounters with God, because God has no sensory or sensuous attributes. In such experiences, what is really happening is that the mind is struck, not by God as such, but by the eternal Forms of the things observed. They are experienced so intensely that the individual properties of their instantiations are forgotten. But God is not the world of Forms, even though He is ultimately the cause of the relationship of our faculties to the formal causes of the world. In any case, the difference between 'illusory' and 'magical' is mainly semantics, and there is no great mileage in it.*

Certainly we do not see God directly; in God's light, we see the Forms or *logoi* of things as symbols of Divine realities. In St. Paul's words (Romans 1:20): 'For the invisible things of Him from the creation of the world are clearly seen, being understood by the things that are made, even His eternal power and Godhead.' And yet, does not the Christian assert that, since the Incarnation, it is actually possible to see God in Christ ('Who has seen Me has seen the Father')? That Christ is the *ikon* of the Father? And that the cosmos, in Christ, is transfigured, until it is all theophany? Maximos the Confessor says:

> He, the undifferentiated, is seen in differentiated things, the simple in the compound. He who has no beginning is seen in the things that must have a beginning; the invisible in the visible; the

intangible in the tangible. Thus he gathers us together in himself, through every object ... enabling us to rise into union with him, as he was dispersed in coming down to us.

My attitude, you suggest, is owing to 'the dark, alienated ways in which abstract thought can construe spiritual truth.' Nevertheless, I construe a lot of other kinds of spiritual truth with the same mind, but with very different results. If conceptual thought indicates something dark and alienated in connection with Non-Dualism, it would be more natural to assume that that is the fault of Non-Dualism; but if we must allow at least the possibility that it is owing to this kind of thought, one still cannot assert that it is so without proof. If one does, it could only mean that Non-Dualism is being put beyond the range of discussion, that is, set up as a dogma.

Abstract thought does not only construe Truth darkly; it can also, within its limitations, open to spiritual Truth; when logically sound it is an inescapable sign of that Truth, as C. S. Lewis so brilliantly demonstrated in his book *Miracles.* Nonetheless, clear, valid logical thought is only a tiny part of our (potential) spiritual experience; most of what we experience—on all levels, not just the spiritual—cannot be expressed in any human language.

One cannot, by definition, logically prove the doctrine of Non-Dualism, since arguing from premise to conclusion is dualistic in essence. Non-Dualism is, and must be, the First Premise as well as the Ultimate Conclusion, and premises, in the sense of axioms, are never arrived at logically; they are understood through direct Intellection. If it were not for axioms witnessed via Intellection, there would be no axioms at all, and therefore no logic. In the case of axiomatic knowledge, proof proceeds from certainty, which is prior to it. It manifests certainty; it does not establish it.

Another reason as to why logical analysis might find something negative about Non-Dualism, from a Christian point of view, is that it requires one to discard a doctrinal truth, i.e., that the world is a real thing, created by a personal God Who transcends it. Having done that, it would not be surprising if the new outlook appeared dismal. It is far easier to be an ex-Christian than to acquire another and commensurate truth. 'The way down is easy,' but the way back up may not be recognizable.

Once having gone down, the way up is totally unrecognizable,

without spiritual Guidance and the virtue of Faith: the evidence
of things not seen. And once again, I do not see Non-Dualism as
asserting the literally illusory nature of the world in every sense.
The introductory essay 'Shankara's Doctrine of Non-Dualism' to
The Crest-Jewel of Discrimination, translated by Swami Prabha-
vananda and Christopher Isherwood, explains it like this:

> When Shankara says that the world of thought and matter is not
> real, he does not mean that it is non-existent. The world-appear-
> ance is and is not. In the state of ignorance (our everyday con-
> sciousness) it is experienced, and it exists as it appears. In the state
> of illumination it is not experienced, and ceases to exist. Shan-
> kara does not regard any experience as non-existent as long as it
> is experienced, but he very naturally draws a distinction between
> the private illusions of the individual and the world-illusion. The
> former he calls *pratibhasika* (illusory) and the latter *vyavaharika*
> (phenomenal). For example, a man's dreams are his private illu-
> sions; when he wakes, they cease. But the universal illusion—the
> illusion of world-phenomena—continues throughout a man's
> whole waking life; unless he becomes aware of the Truth
> through knowledge of Brahman. Shankara makes, also, a further
> distinction between these two kinds of illusion and those ideas
> which are altogether unreal and imaginary, which represent a
> total impossibility or a flat contradiction in terms—such as the
> son of a barren woman.

> Here, then, we are confronted by a paradox—the world is and it
> is not. It is neither real nor non-existent. Any yet this apparent
> paradox is simply a statement of fact—a fact which Shankara calls
> Maya. This Maya, this world-appearance, has its basis in Brahman,
> the eternal. The concept of Maya applies only to the phenomenal
> world, which, according to Shankara, consists of names and
> forms. It is not non-existent, yet it differs from the Reality, the
> Brahman, upon which it depends for its existence. It is not real,
> since it disappears in the light of knowledge of its eternal basis.
> World-appearance is Maya; the self, the Atman, alone is real.

I would add one further dialectical step to Shankara's (or the
translators') argument: that the world-appearance both disappears and does not disappear in the light of the knowledge
of Brahman. In the *nirvikalpa-samadhi* it disappears; in the state
of waking consciousness of a sage like Ramana Maharshi it

obviously reappears, but is directly recognized as illusory in itself and not other than Brahman in its essence. Ramana Maharshi could recognize individuals, respond to specific questions, walk deliberately from point A to point B; his perception was not absorbed in an undifferentiated field of Divine Light. Yet the disappearance of the world-illusion that he had already experienced, and undoubtedly often returned to, was always there in the background of his day-to-day experience of the world-appearance; it was this very realization which revealed that appearance to be illusory.

There is a problem with Indian thought here, which is not just mine. Europeans are spiritually Egyptians—the Judaeo-Christian tradition, Platonism, Pythagoreanism, and Hermeticism are all of Egyptian origin. The fact that God acted in history through Moses and in Christ clearly did not alter its essential nature. Therefore, trying to graft Indian spirituality onto people with that heritage is really just a way of increasing the amount of confusion in the world.

There was an age, before the tower fell (or before the vision of the Presence of God fell, after which man was subject to the Promethean temptation to 'try and put Humpty-Dumpty together again' by building such a tower), when 'India' and 'Egypt' were one. In the spiritual Heart of man, that age is with us still.

Sincerely,
Charles Upton

[NOTE: Below I have inserted my responses to Robert Bolton in Roman type in the body of his letter.—C.U.]

Dear Charles Upton

In reply to your question about Schuon, I do not remember where that passage was, but only that Schuon was speaking conditionally, as in fact I was: given a certain idea of the human condition, then must follow absurdity etc. Granted that Schuon was using the word 'dream' metaphorically in this context, the implication of this metaphor, 'a world created out of one's own substance', is that the world is not a result of creation but of emanation.

What happens in dreams has no more relation to our conscious intentional-
ity than the shadows cast by our bodies. Such a conception clears the way
for a mysticism of identification, but the word 'creation' in this context
could only be rhetorical.

Certainly the metaphor of 'dream' has more to do with emanation than creation. Nevertheless it is my belief that everything
God does is both intentional and effortless. The 'Divine Workman' metaphor refers to the first aspect, the 'dream' metaphor
to the second.

You say further on that my arguments about the collective dream 'deny
the universal testimony of the mystics,' rather as though there were only
one such testimony, and that perfectly clear. What happened to St. Paul
does not amount to a universal testimony. The great majority of mystics,
in Catholic tradition at least, did not take their experiences as a basis for
changing their ideas of reality, but were rather reluctant even to speak of
them. This was because their experiences gave them a new dimension to
what was contained in their faith in any case. St. Benedict and St. Teresa
of Avila were typical in this respect. One could only say I was denying
their testimony by drawing conclusions from their visions which they did
not draw.

I agree that the traditional mystics who speak of 'awakening'
are not talking about an experience that invalidates their doctrinal knowledge, but represents the realization of it, the passage
from belief to direct witnessing. This is the experience spoken
of in II Peter as 'the daystar arising in one's heart.' In the words of
Jesus, 'I come not to destroy, but to fulfill.'

On the question of seeing God in nature, it is true that we do no perceive
the Forms exclusively, because God relates to all the Forms as each one of
them does to all its instantiations, so God is seen in an implicit way in every
Form. A Unity which contains too much for human minds can be known in
a divided and serial manner. However, I do not follow the transition from
the Incarnation's manifestation of God the Father to His being manifested
by all nature as well. If God were manifest to that extent, would the Incar-
nation have been necessary?

I was referring to the Orthodox Christian doctrine that
Christ, by His Incarnation, has transfigured the natural world,
transforming it (for those who participate in the mystery of His

Incarnation and Atonement) from a fallen reality (a 'veil') into a theophany—which is what it was in Eden, and is in essence.

Concerning proof of Non-Dualism, you say that it must be 'the First Premise as well as the Ultimate Conclusion,' and that it is an axiom 'witnessed via intellection'. However, if you wish to claim that, you must show what your axiom actually states. From what you say, you believe its truth to be self-evident, which looks highly unlikely, because there is no agreement as to whether even so basic a statement about God as 'God is' is self-evident. In act, it can only be so if one accepts Anselm's Proof, and that in turn requires Platonic premises.

On the other hand, it is easy enough to see one basic proposition for Non-Dualism, namely, 'God is, therefore nothing else is,' but there is no self-evidence there. If true, it would mean that 'God is the Creator' was self-contradictory, which logicians have never suspected it of being. It is also hardly self-evident that the self and God each consist of two separate things: the one, a metaphysical part and a collection of psycho-biological junk, and the other an undetermined Absolute and a personal Creator; and that the two metaphysical parts fuse while Maya disposes of the biological junk and its Creator.

I say that God IS (though He is not limited to this 'isness'), while everything else both is and is not: everything IS as a manifestation of Divine Reality, but IS NOT in terms of its creatural limitations. Nonetheless, these limitations are necessary for the appearance of creatures—an appearance which God wills. So what are we to make of this?

This is not to deny a strong form of psychological self-evidence in this connection, which is true in general, even if not of you in particular. If one takes the position that there is no creation; no one to pray to; no redemption; no immortal soul; and no eternal life for human beings as such, one has given up a lot (and even more if one had some belief in them in the first place). That kind of asceticism creates an unshakable sense of moral superiority over those who do not follow this path, and this combines powerfully with the mind of the unconverted and despairing pagan who remains somewhere inside even the devoutest of us. He always thought that life came from nothing and returned to nothing, and now he can come back home to stay, with a spiritual role to play. No one can deny the strength of this combination.

Indeed. The 'asceticism' you describe can only be that of the *awliya al-Shaytan*, Satan's contemplatives. The 'mysticism' which denies the soul, the soul's redemption, eternal life and a merciful God to Whom one can always turn, in all suffering, or out of the need to praise and give glory, is not mysticism but nihilism. Shankara himself composed beautiful hymns of praise to God, which can only mean that the transcendence of Ishvara in no way entails the denial of Him.

This underlines the fact that Non-Dualism does not require any faith, such as one must have to believe in personal salvation. It also begs the question that we might, after all, be created, and created specifically for the purpose of finding individual salvation and immortality in Heaven. (I know Non-Dualists believe that Heaven is only for stupid people, but if they are so intelligent, one would expect them to be able to see that this opinion of theirs results only from an idea of Heaven they acquired in childhood and never developed further.)

I remember once explaining Guénon's Non-Dualist gospel to someone who knew nothing about it, at a time when I still believed in it. To my amazement, he said it was hardly any different from Communism! But later on, I was not so amazed, when I saw how both systems attack people through their moral sense with a cult of self-elimination, where they find the idea of personal redemption completely incredible. In either case, this despair gives rise to a rejection of individuality as such, and with it any spirituality of creativity.

Your experience confirms so much of my own when it comes to certain 'Traditionalists', though not to the essence of Traditionalist doctrine. The egotism based on a tendency to despise all that is created and all that is personal—and which is in fact scandalized by the least mention of love, whether human or divine—is certainly the worst form of egotism I can conceive of. 'Blessed is he who is not scandalized in Me.'

Concerning what you say about reality and illusion, and the text you quoted about it, it is not yet clear to me how far that coincides with what Plato and Aristotle have said about the way in which nature falls short of the reality of the Forms instantiated in it. To try and see how near or far apart we may be on that subject had best wait till another time.

A general point about Non-Dualism is that, if it were true, one would expect that finding God and finding one's own real self would be one and the same thing, whereas real life shows again and again that they are not. It is possible to find God in a sad and frustrated kind of way without finding one's real self, and those who go too far down that path end up as the 'holy idiot' types found in religious cultures. Conversely, those who find their real self often fall into the trap of thinking that they must have found God as well. Possibly Gurdjieff did that. At any rate, his usefulness is mainly for those whose need is to find their real self.

I assume the 'self' you are speaking of here is the psychic self, the soul. The Self the Advaita Vedantins speak of is the spiritual Self, the Atman, which is closer to what Christians mean by 'synteresis' or 'Nous', though the two may or may not be strictly identifiable. (Most Christians, I'm sure, would not strictly identify them.) You touch here on the all-too-common tendency of those attempting to follow a spiritual Path to use that Path as an excuse not to understand themselves in human, creaturely terms, and then to justify this flight from self-understanding by pontificating as to how the Atman or Nous so vastly transcends the 'mere' soul or psyche. But it is this 'mere' soul that is saved or damned, which means that we had better understand ourselves on this 'all-too-human' level as well as we possibly can. The fact that the Atman is undamnable and eternally one with God, and that we are host to it intrinsically, is a 'mere' truism which has NOTHING to do with the salvation or damnation of our immortal soul. If we are so involved in the passions—chief of which is the passion of pride, whose metaphysical expression is the error of solipsism—that we are set to inherit Hell, then there is no way we can realize the moksha of the Advaitins. And to use the mere mental knowledge, the mere belief, that 'my truest "I" is God' as an excuse to not even BEGIN the path of self-purification, the 'unseen warfare' against the passions for the purpose of realizing the virtues—which absolutely REQUIRES human self-understanding, as well as all conceivable humility and gratitude to God for His grace and help, without which we would be irretrievably lost—is nothing short of sacrilegious. It is also inconceivably foolish. To think that you can read some books on the Vedanta, understand them well enough mentally so that you can expound them, do maybe a few meditation exercises, and thereby become a realized Non-Dualist Sage—without ever having confronted the spiritual darkness in your own soul, not to mention having gone through the metanoia that such a confrontation calls for but cannot by itself accomplish—is—well, it's just plain dumb. 'Paper cakes do not satisfy hunger.' This is the ingrained foolishness of so many of these 'Traditionalist' wise men. Thank God I have been privileged to meet certain other Traditionalists who have not been deluded by ideas. I am thinking of one person in particular, a saint in the making, the most canny and unglorious of men, whose presence nonetheless radiates, in spite of himself, the Uncreated Light. [Sincerely, Charles Upton]

Yours sincerely,
Robert Bolton

Dear Robert Bolton

One thing stuck in my mind from our last dialogue. You had objected that the Vedanta denies personal immortality, and I had agreed that to deny such immortality in a simple way is no part of true religion. Since then my wife and I visited Dr. Coomaraswamy, and I put the question to him: Does the Vedanta deny personal immortality? His answer was: 'Ultimately, yes.'

That got me thinking. Now it is clear that Hinduism as a whole does not deny personal immortality in the rudimentary sense of 'the survival by human consciousness of bodily death.' And I still do not accept that the Vedanta is a sort of hybrid of an abstract Absolute with the kind of naïve realism that sees in birth and death only the appearance and disappearance of bodies—this, according to you (if I understand correctly) being the genesis of the theory of reincarnation. Hinduism posits many after-death states that go far beyond simple 'physical' reincarnation: hells, paradises of form, paradises beyond form, *lokas* of all descriptions. The real question, then, is not 'does human consciousness—in some form or other—survive bodily death?'—clearly Hinduism answers in the affirmative—but rather: Am I, in my present human individuality, immortal? Does Charles Upton—presuming that he is not immediately annihilated in Brahman—change into someone or something else at death, or does he remain Charles Upton, himself and no-one else, in heaven, purgatory or hell?

A child might ask: When I am an adult, will I still be myself? The answer is: yes and no. An adult is no longer a child—yet if his or her adulthood is a true maturation, a full and balanced fulfillment of latent possibilities, then it is clear that whoever the child was still lives on in the adult—not as the proverbial 'inner child', but as the intimation or fore- shadowing of a fuller reality now realized. Potency asks: When I am actualized, will I still exist? The answer, again, is yes and no: Potency lives on in Act, yet for this Act to be completed, potency—as potency—must die. He who loses his life for My sake—for the sake of greater Life—will find it, but he who seeks to retain his life—who wishes to remain always a child—will lose it. He will be no adult, but neither will he any longer be a child. In trying to retain the person he once believed himself to be, in refusing to die into the Greater Life, he will lose that Greater Life—and the lesser one too.

An actor enacts the part of Hamlet. He gives a certain degree of life and reality to that role; he may even forget, for the duration of the play, that he is not really Hamlet, but an actor portraying him. But when the curtain comes down, Hamlet dies: he was such stuff as dreams are made on. All that Hamlet was, was drawn from the soul of the actor. The play 'Hamlet' by William Shakespeare may have determined what particular set of potentials would be drawn out of that soul to give fleeting life to the title character, but there was nothing in that characterization which was not first in the soul of the actor enacting it.

In the case of the child who becomes an adult, and in that of the actor who finishes with the play, neither the child nor the character are 'cut off' or 'left behind': they have simply returned to their archetypes. (Plotinus says that the one who turns his face away from a mirror does not leave his image imprisoned in that mirror, cut off from the reality of his face: the image simply ceases to be, and nothing whatever is lost, since the face remains.) The child was the hope that the complete man might come—this is very the charm and genius of childhood, of all that is signified and foreshadowed by 'play'—but the man is the fulfillment of that hope, and when the fulfillment comes, the hope and longing for that fulfillment no longer have a legitimate place. Now that the play is ended, the character Hamlet has been revealed as a partial aspect of the real man. And even though Hamlet, as a separate character, is now no more, all that Hamlet was remains integral to the soul of the actor who played him, a soul which contains infinitely more reality than any theatrical portrayal could encompass. Nothing is lost, everything is gained, when the curtain comes down. In terms of the Advaita Vedanta: Brahman is the sole Transmigrant, the one and only Player. As Rumi puts it (in my paraphrase): 'I died as mineral and became plant, died as plant and became animal, died as animal and became man. When I die as man, I will be angel; when I die as angel, then I will be the Inconceivable. What have I ever lost by dying?'

Earlier, you said: *For the Indian doctrine, man is something which is really the same as God, but which gets all kinds of cosmic pollution stuck to it in the course of arriving in this world. So, then, we just have to scrape off the pollution, and there will then be nothing but God, just as it ought to be. This is practically the same as saying that man as such is not real at all.* But, following Plotinus' metaphor of the mirror, the mirror-image of the face is not 'cosmic pollution' which much be 'scraped off'—such an idea is really more Gnostic (in the sectarian sense) than Indian—but is simply a 'special case' of the reality of the face in a dimension of lesser reality than that intrinsic to the face itself. The mirror-image is not a pure illusion; it indicates the face and is derived from the face. And it is really there. It is not a subjective hallucination. Anyone approaching from the requisite angle a mirror set opposite a face will see the image of that face; the face need not be mine, it could be yours just as well. Yet the image is not the face itself: it is a reflection, a product of *maya*. And when the image is dispelled, it does not remain lying on the ground as a skin or shell of cosmic pollution scraped off the reality of the Essence. It is gone. It was—and yet, in another and equally true sense, it never was. It was: this is its *suchness.* Yet it never was: this is its *voidness.* Such is the nature of *maya*: it is and it is not. If it possessed being exclusively, then appearance would be God, and we would all perforce be pantheists. If it possessed no being, it would not appear, and God the unknowable would have no way of being known, nor would you or I or anyone else be here to know Him.

God the Father leads us on from mansion to mansion of His infinite House. 'Men are asleep, and when they die, they awaken' said the Prophet Muhammad, peace and blessings be upon him. The part awakens into the whole, to become the whole, to see that, in a very real sense, it always was that whole—a whole which, in turn, is only part of an even greater whole, on and on up the Great Chain of Being, until *samyak sambodhi,* the final Awakening. And this ascent of Jacob's Ladder is in no way a melting away into abstraction, but is rather the realization of ever higher orders of particularity and concreteness, which is why it is said that 'every angel is a species as well as an individual.' An angel is not one individual of a species, but each angel *is* the species itself. And he is not deficient in individuality by virtue of being identical with his species; on the contrary, he possesses individuality superabundantly. And when the level of the Divine is reached, then we can truly say that God is the *only* Species, the *only* Individual: 'Before Abraham came to be, I Am.' Abstraction is the elimination of any particulars that would place an entity outside a given category; it seeks a lowest common denominator. Ontological ascent is a gathering together and synthesis of particulars in a greater Particularity. In Platonic terms, the Idea of the Horse cannot be an abstraction (though our *language* tends to become more abstract as we discourse on higher particularities—which can clearly be seen, for example, in the three books of the *Divine Comedy.* The fact that discursive language becomes more abstract as its object becomes more concrete has fooled a lot of people.) If everything were eliminated from the Idea of Horseness which could differentiate one particular horse from another—color, gender, size, age, health, intelligence—who could recognize this shapeless blob of 'horseness' as a horse? The true Idea of the Horse must be a super-saturated reality which encompasses all the particulars of all actual horses, and perhaps all possible ones, in a higher synthesis. It is invisible to our senses not by deficiency of reality, but because those senses are not designed to encompass such richness, being relatively abstract in relation to the Intelligibles, which are more *concrete* than sense-experience. (Our senses, being 'organs of limitation', work by abstracting sense-experience out of the infinite Concreteness which surrounds and interpenetrates and composes us: 'Five windows light the caverned man,' said Blake, commenting on the Allegory of the Cave.)

When I return to my Archetype, my Idea, I am more truly myself than when I identified myself with, and limited myself to, a sensual or psychic abstraction drawn from my fuller, integral, eternal nature. When Ramana Maharshi directs us to ask ourselves, 'Who am I?,' he is leading us on toward that fuller nature, toward the realization of ever greater levels of concreteness and essentiality. Toward immortality, if you will.

There is, however, another way of looking at personal immortality, which I have already intimated above.

As you have pretty much pointed out in terms of Christianity vs. the Vedanta, the western, Abrahamic spiritualities—Judaism, Christianity and Islam—tend to emphasize personal immortality in their accounts of the ultimate destiny of the soul, while eastern religions, notably Hinduism and Buddhism, see the persistence of individual identity as something to be liberated from, not something to seek as if it were the highest goal of the spiritual life. Even the most exalted Hindu paradises are still part of the realm of name and form, and thus subject to mutability and decay. When the soul exhausts its favorable *karma*, or when the entire formal universe itself is reabsorbed into the Formless Absolute, the personal immortality of even those souls who have attained posthumous bliss is ended. To say it another way, Liberation is virtual within the western religions, while immortality is explicit; within the eastern religions, the reverse is the case.

Are the notions of personal immortality and ultimate Liberation therefore fundamentally opposed? I don't believe so.

As time-bound, contingent beings, we are mortal; as the Buddhists never tire of pointing out, no relative and compounded state, either incarnate or posthumous, can be eternal. In the words of the Qur'an, 'All is perishing except His face.' Yet God *is* eternal, and as such He sees all things *sub specie aeternitatis*. What to us is a temporal event is to Him an eternal form. In time we pass and die, and are reborn in higher or lower worlds, only to pass and die again. But if ever we once were, in God we eternally are; this, not some indefinite persistence through time of our relative, compounded selfhoods, is personal immortality. Yet this immortality is not a property of ourselves as self-reflexive subjects, but of God, or rather of our eternal archetypes in God. We are immortal, in other words, not in terms of our experience of ourselves as continuously existing beings, even in blissful communion with God, but in terms of *God's experience of us* as reflections or aspects of Himself.

Hinduism posits three planes of being: *kama, rupa,* and *arupa. Kama*, as 'desire', is the principle of time, which in terms of the consciousness of sentient beings is based on the desire to possess or avoid this or that experience. *Rupa* is 'form', and it transcends time; that which is perfectly itself doesn't need to seek or avoid something in order to complete itself or protect itself. And *arupa*, which transcends *rupa*, is the Formless. God is Formless, yet within Him is form, to which He lends, not His self-sufficiency, but His immortality, or a certain degree of it. The world of timeless forms or Platonic archetypes is not self-subsisting; it depends for its existence upon the Formless Absolute. Nonetheless, it still transcends duration as we experience it. Even though it does not partake of the perfect eternity of God, it remains 'relatively eternal' in relation to passing time.

As a reflection of God's eternity, *rupa* is equivalent to the Greek concept of *aion*. 'Aeonian time' is a portion of passing time considered as a single eternal form, much as a year can be represented visually on a single page as a cycle of twelve months. To say 'time is the moving image of eternity' (cf. Plato's *Timaeus*) is to posit aeonian time; this is the level of personal immortality. What in passing time is a lifetime of experiences and choices, in aeonian time is a single eternal form, enthroning a single, complex, sovereign choice, a timeless choice like that of the angels, who chose to obey or disobey God 'before time began'. But because of the nature of eternity, which underlies the moments of our lives rather than simply coming before or after them, this eternal choice does not strictly predetermine our daily choices in passing time, since it can with equal validity be considered as the final sum of them. (This, incidentally, is the metaphysical, esoteric meaning of the Calvinist doctrine of pre-destination, which when interpreted exoterically—as it always is—posits a capricious God married to hopeless fatality.)

Just as time (*kama*) is a moving image of the eternal forms (*rupa*), so these forms are dependent upon the Formless (*arupa*), without which they would be caught in the stream of passing time, and therefore not be eternal. In Buddhist terms, the fact that the Formless is the source of all forms is the principle of the 'voidness' of forms, their absence of self-nature; that which has no self-nature can neither persist nor pass away.

We are immortal in our limited, human forms because, due to God's vision of all forms *sub specie aeternitatis*, whatever has once existed necessarily exists eternally; we are not simply subsumed or dissolved in the Divine Nature. Yet this immortality is real only by virtue of our total Liberation from self-reference and the illusion of self-existence, the factors upon which time and mortality are based—a Liberation which seems, from the temporal point-of-view, like a goal to be attained, but which from the standpoint of God's eternity is simply the real nature of things. As the Cha'an Buddhists say, 'from the beginning, not a thing is.'

If this eternally-existing Liberation is not realized, we seem to wander from form to form in passing time, lost in the illusion of *sangsara*, always dying yet never cleanly annihilated, always becoming yet never quite able to be. This *sangsara* is based on our craving to *be our own experience of ourselves*—a craving which can never be satiated, since an experience can never be objective to itself, and thus can never be fully experienced; it remains a mere virtual reality. Our immortality as limited, contingent forms cannot be based on our own experience of ourselves, but only on God's experience of us as eternal reflections or aspects of Himself, according to which—from one point of view—He is the only Experiencer, yet equally no Exeriencer at all, since in essence there is nothing other than That One for Him to experience, and to 'experience' oneself on a level transcending the subject/object polarity is to transcend experience itself. The eternal reality of a human form is only realized when that form dies to itself, when the ego is annihilated and the Eternal Divine Witness, the *atman*, is unveiled; this is one meaning of the *hadith* of Muhammad (peace and blessings upon him), 'he who knows himself knows his Lord.' This is *moksha*, Liberation. Personal immortality, in other words, exists only by virtue of the total objectification of the human subject before God, the annihilation of its illusory self-nature, which is complete liberation from ego. Where self-identification is ended, the human self remains as one among God's infinite possible Self-experiences—He who, in another sense, is absolutely beyond experience, since He is 'One without a second'. This is the *fana* and *baqa*, the 'annihilation and subsistence' of the Sufis. It is the import of those Buddhist *thankas* (roughly equivalent to icons) where this or that enlightened sage is represented in terms of his or her eternal form, empty of self-nature yet available to the not-yet-liberated as an intercessor, a channel for the transmission—virtually at least—of Perfect Total Enlightenment. It is what St. Paul meant when he said, 'it is not I who live, but Christ lives in me,' and the real import of the words of Jesus, 'he who seeks to keep his life shall lose it'—i.e., the attempt to exist as one's experience of oneself leads to endless wandering in *sangsara*—'but he who loses his life, for My sake'—for the sake of the *atman*, the indwelling Divine Witness—'shall find it.'

There is no essential contradiction, then, between self-annihilation, personal immortality, and ultimate Liberation from the round of becoming, these being three different aspects of the same Reality. Approaches differ; the Truth is One.

When I wrote the above passages asserting that there is no ultimate incompatibility between personal immortality and Liberation in the Supreme Identity [See *Personal Immortality vs. Liberation* below, p239], I was speculating (for all I knew) alone. Since then I came across the following passage from an article by Martin Lings in *Studies in Comparative Religion, Summer-Autumn* 1979, entitled 'Sufi Answers to Questions on Ultimate Reality':

> . . . . every pious believer may expect for himself or herself not only one Paradise but two [the Garden of the Essence and the Garden of the Spirit]. The Garden of the Essence is no less than the Absolute and Infinite Oneness of God. From this point of view it might seem that all other Paradises cease to exist. How then can it be said that the supreme Saint, who is by definition in the highest Paradise, has 'also' a second Paradise? This statement can be parried with another: If it is possible for a supreme Saint to say, during his life on earth, in all sincerity of gnosis, 'I am the Truth,' why should it not be possible for such a statement to be made eventually, by the same Saint or another, in the penultimate paradise, the Garden of the Spirit?

> This brings us . . . to Schuon's *Islam and the Perennial Philosophy*. . . . In the final chapter, 'The Two Paradises', he reminds us that 'there are in man two subjects—or two subjectivities—with no common measure and with opposite tendencies, even though there is also coincidence between them in a certain sense.' The Divine and the human natures of Jesus and their equivalent in Muhammad are ideal examples; and if it were not possible for the two subjectivities to co-exist, albeit at different levels of reality, in the next world, then the Prophets and the Messengers, once they had departed this life and been integrated into the Paradise of the Essence, would be altogether withdrawn from existence as differentiated persons, and the possibility of contact with them would be irretrievably lost. As Schuon says: 'we should have to conclude that the Avatara had totally disappeared from the cosmos, and this has never been admitted in traditional doctrine. Christ 'is God' but this in no way prevents him from saying 'Today thou shalt be with me in Paradise,' or from predicting his own return at the end of the cycle. . . .

> The Sufi conception of our final ends certainly allows for the duality in question. Rabi'ah al-'Adawiyyah's utterance of the adage 'the neighbor first, then the house' in the sense of 'God before Paradise' . . affirms not only a precedence but also a co-existence; and it may well be asked if there has ever been a Sufi who did not hope to see the Prophets in Paradise, despite such

formulations—more methodical than doctrinal—as have misled some Western scholars into supposing that what the Sufis aim at as regards their individualities, is 'blank infinite negation'. Christ's 'Seek ye first the Kingdom of Heaven, and all the rest shall be added unto you expresses a universal principle which dominated every mysticism and which is often, as in Sufism, transposed to the highest possible level to mean, by 'the Kingdom of Heaven', no less than the Garden of the Essence . . . if a Sufi were asked 'What is Ultimate Reality?', let us suppose that his answer 'the Divine Essence', calls forth the objection: 'But I mean your Ultimate Reality'. His answer to this might well be: 'The beginning and end of my subjectivity are in the very Self of the Divine Essence.' And if it were still further objected: 'But I mean *you* as distinct from anyone else,' he could insist: 'You cannot escape from the Divine Essence as answer, for Ultimate Reality is One. That which you ask of is there, among the immutable essences (*al-a'yan ath-thabitah*) which are the supreme archetypes of all differentiation, mysteriously united in the Oneness of the Self.' But a possible answer to the last question would be 'the Paradise of the Spirit'. That is the summit of what is normally understood by Paradise; and though this answer is not as rigorously metaphysical as the first, it may also nonetheless be acceptable to Ultimate Reality Itself, which has ordained that Paradise shall be 'not other' than the Ultimate.

To 'die before you are made to die' is to realize the Paradise of the Essence, the Supreme Identity of the Atman, in this very life; this is what the Sufis mean by *fana*. And it is this very realization of the Absolute Witness which reveals the immutable essences that constitute what Schuon calls 'maya-in-divinis', among which are the *aions* known as Robert Bolton and Charles Upton. This is *baqa*. Thus the Realization of the Supreme Identity is precisely the seal and guarantor of personal immortality. If our 'second subjectivity', the psycho-physical one, becomes totally objectified before the face of the 'first subjectivity', the Absolute Witness, then even its own conscious, reflexive self-knowledge can subsist, by virtue of the seemingly paradoxical fact that Robert Bolton and Charles Upton are no longer identified with it. It is no longer a veil covering the Supreme Identity, but a direct manifestation of It. This is the resurrection of the dead.

When attending the deaths of my mother and my best friend, I noticed an interesting thing: as the end approached, before consciousness was lost, both began speaking of themselves in the third person. When I asked my mother 'How are you feeling?' her answer was, 'She's alright.' The dawning of the Absolute Witness—or at least the momentary flash of it which, according to the *Bardo Thodol*, comes to all at the moment of death—had begun.

Dear Charles Upton

In all you have said, you have not addressed the basic point I made earlier, that all this panorama of progressive depersonalizations can be taken, without challenging the facts, to be changes going on among the contents of the soul of one person. It is a collection of thoughts, and you know them, but they do not know you; to suppose that you yourself could somehow be subsumed into your own thoughts and ideas is as absurd as to think that you will get knocked down by your own shadow. Such attempts to get away from personhood are as futile as Monkey's attempt to jump out of the hand of the Buddha.

I am infinitely more than any system of thoughts that I may choose for myself or have offered to me, because my personal being is the primary reality, not my ideas. Hence personal immortality, and hence also a charitable understanding of the destiny of other people.

Yours sincerely,
Robert Bolton

Dear Robert Bolton

You say: *I am infinitely more than any system of thoughts that I may choose for myself or have offered to me, because my personal being is the primary reality, not my ideas.* I counter: God, not my personal being, is the primary Reality, and God is not one of my ideas. In order to use language, I must draw upon ideas which are 'mine' only insofar as and only so long as I employ them, but God in His Reality is not one of those ideas. Rather, He is That to which all ideas ultimately refer. Neither my personal being nor yours is the primary reality, because God is Reality itself.

I have only one more question: You say that my ideas are totally dependent upon my own personhood, that I am aware of them but they are not aware of me. But since you call yourself a Platonist, don't you therefore hold to the doctrine of a plane of being where Ideas exist in their own right?

Sincerely,
Charles Upton

Dear Charles Upton

I was not saying that my thoughts create Forms—they have an independent subsistent reality of their own, but they (or most of them) are subordinate to souls in the Chain of Being. I think Ficino understood it that way.

Yours Sincerely,
Robert Bolton

Dear Robert Bolton

Of course the 'ideas' I entertain minute-to-minute cannot be identified with the Platonic Forms. They are too mired in my own subjectivity to be so identified. Yet any clear conception of Truth is a product of the 'grace' emanating from those Forms. The one thing I want to establish, after all this yammering, is that there is something IN us that is GREATER than us, and it is this that saves our personhood from being no more than a self-enclosed ego—a kind of meaningless impurity that must, in your words, be 'scraped off' the Absolute in the course of the spiritual Path.

Sincerely,
Charles Upton

Is Buddhism
Really Atheism?

A Letter to Gary Snyder

after Phil Whalen's Memorial Service,
Green Gulch Zen Center

Dear Gary,

You told me, at Phil Whalen's memorial service, that Buddhism
is really atheism. Is this actually true? (It all depends, of course,
on what we mean by 'God'.)

Buddhism has plenty of names for the Supreme Principle:
Nirvana, the Buddha-nature, the Dharmakaya, the Adi-Buddha,
the Clear Light of the Void...it also recognizes that 'to name
something is to kill it,' that whatever mental images we make of
this Principle are nothing but obscurations. This is why Buddhists are very wary about speaking in terms of a Principle at
all—though, in fact, they always have.

But are the names of the Supreme Principle always obscurations? Are they not also, sometimes, *upayas*, like all those bells
and candles and statues at Green Gulch? When Shakyamuni
reached perfect total enlightenment, he had to face this question. He knew that if he spoke, anything he said would be a distortion, but that if he didn't speak, he would be a great *pratyeka-
buddha*, but no teacher, no savior. Brahma had to appear and beg
him to teach. Brahma represents the first dawning of the obscu-
ration/teaching itself; the way he dawned and asked to be taught
recapitulated the 12-linked chain of causation (on the way out)
but also the 8-spoked wheel of the Dharma (on the way back).

We have to speak. We have to say, 'there is, monks, a realm devoid of earth, air, fire and water. It is not the plane of infinite space, nor the plane of infinite ether, nor the plane of infinite consciousness....it is the ending of sorrow.' We have to say, 'if it were not for the Unborn, the Unmade, the Uncompounded, there would be no liberation from what is born, made, compounded.'

IT. IT is unborn, unmade, uncompounded. And IT is no different from Eckhart's Godhead, or the God Beyond Being of Dionysius the Areopagite (whose *Mystical Theology* reads just like a Mahayana sutra), or the Divine Essence, the *Dhat* of the Sufis . . . which, of course, is not an IT, if an IT is a SOMETHING over there with a ME sitting here looking at it.

The forms of all things arise. Among them are the forms conducive to Perfect Total Enlightenment: The human form itself, and the various forms of the Dharma which dawn to teach it its true nature: the sutras, the human buddhas, the celestial bodhisattvas.

All these *upayas*, all these saving-things, are 'nothing but' mental images or constructions. There is no 'Buddha'; he is only a form in the mind—and the Buddha is the Buddha because he holds no such form in the mind. There is no 'God'. He is only a form in the mind, and God is God because He is not limited by the idea that He is God—in Ibn al-'Arabi's terms, He is 'absolutely non-delimited—that is, not delimited even by His own non-delimitation.' (NOTE: The best introduction by far to the teachings of Ibn al-'Arabi, the Shaykh al-Akbar, the Greatest Sufi Shaykh, is *Imaginal Worlds* by William Chittick.) All these 'things' I've been talking about are merely images arising in the Mind.

The hooker, of course, is the 'merely'. When Phil Whalen, on his long death-bed, said of his hallucinations, 'I know that these are only images arising in the mind—but *damn!*', he was speaking from the idea that all these images are arising in 'my' mind: the mind of Phil Whalen, or Gary Snyder, or Charles Upton. At that moment he was not living in the realization that the Mind in which all things arise and fade away is 'mine' only because it is the real nature of what I mistakenly call 'my' mind—that in this Mind 'my' so-called mind, and this universe we seem to be a part of, including this very body, arise interdependently. Like Burroughs once said: 'Maya, am I? You don't get rid of me that easy.' Phil thought he was hallucinating; he forgot that it was the Void that was hallucinating, and that he was one of the hallucinations—and that hallucinations don't hallucinate.

All things arise and fade away not in my mind, but in The Mind, not in this little ego-self, but in that which we Children of Abraham—and we Children of Manu—call 'God', Who is absolutely non-delimited, Who is 'without self-nature'.

Does this 'Mind' then, really exist? Iamblichus called it 'Beyond Being'; does something that is beyond being really exist? It exists in the sense that, void though It may be, it generates versions of Itself which have the power to take us beyond ourselves, to save us from ourselves, to liberate us. Dharmakaya emanates Sambhogakaya, the apparitional Buddha, which in turn emanates Nirmanakaya, the enlightened human being. All this arises and fades away in the Mind, which, since it is the object of no knowledge because it is not known by any knower, is no object, and in that sense No Mind. Yet we do really hope—do we not?— that Phil Whalen could jump through that turbine engine of that jet plane, like Dick Baker told him to, and come out perfectly shredded, at one with the Dharmakaya. [NOTE: This is a reference to the 'guidance through the after-death *bardo*' teaching spontaneously given by Richard Baker Roshi at the memorial service, in view of the fact that Phil Whalen had been an airplane mechanic in WWII.]

If not, it was a pity to say so.

It is useful *upaya* for those who believe in and hold to concepts as if they were rocks or cars to say: 'the dharma-body is void.' It is also useful *upaya* for those who think that the world is nothing but a dream in their own little narcissistic minds to say that we must 'seek union with what goes on whether I look at it or not' (Lew Welch). If Perfect Total Enlightenment is not the real nature of things, if all of us are not enlightened from the beginning, if all things are not Buddha-things, then we are irretrievably lost. In the morning lecture on the day of the funeral, somebody asked the nun who was speaking, 'it sounds like you are saying we should just do right, just be good, just practice virtue. Is it as simple as that?' 'It's very hard for sentient beings to change,' she replied.

I disagree. I don't think it's hard; I think it's impossible. If Perfect Total Enlightenment doesn't go on whether we look at it or not, if it isn't always there to turn to, always there to compassionately save us, then we are thoroughly screwed. A karma-conditioned, karma-created being trying to perform Right Action without generating more karma? No Way. That is not the Way. There is no Way there. Action can be realized and performed as a *paramita* only because everything has already been perfectly done—emptily done—done without any doing at all: *wu wei*. It's already there. It's already over. It's done. What's done keeps on getting done only because it's already over. Empty of self-nature. *Finito.* (The Christians say the same when they say 'Our debt is infinite, yet Christ has paid it all.' The Buddhist equivalent is 'sentient beings are numberless; I vow to save them all,' which is just another way of saying 'the Buddha holds no such view as 'numberless sentient beings in need of salvation,' and therefore 'all beings are enlightened from the beginning.')

But if it isn't already done, if we find ourselves bound hand and foot in the Hell of Self-Improvement, tormented by demons of spiritual aspiration, then we are without hope. *'Work—work—work—work—work'* (to quote one of Phil's poems)—we'll never get it done, we'll never get it paid; karma is the coal-mine with the company store. Self-power, if the self is the ego, is powerless. The ego provides no way out of the vicious circle (*'samsara'*) of the ego. Other-power is needed if self-power is to have any power—the power of something other than the ego: that's why Buddhists take refuge in the Buddha the Dharma and the Sangha. Self-power is the vow to walk according to other-power—the power of something other than the ego. And if there is no ego—when there is no ego—given that there is no ego—*jiriki* and *tariki* are one. (While I was writing this, the Fed Ex man came to the door and left a box full of wonderful books, unexpected gifts from a friend in Indiana, among which was *Naturalness: A Classic of Shin Buddhism*, by Kenryo Kanamatsu!)

The Truth is Always So; it goes on whether we look at it or not; and for that reason, we can turn to It; and for that reason, It can save us. *Upayas* are not arbitrary or contrived; they are in the nature of things; they are modes of Wisdom. (As Chögyam Trungpa once said, 'effort just comes to you.') But if we think we somehow have to *invent* It first and then make It work—God help us.

American Buddhism is sometimes nothing but a way for materialists—those 'primates' you mentioned—to seem to practice religion while still being good materialists and secular humanists. I've always wondered if that's why there are so many Jews in it: they are in flight from the Lord God Jehovah, not realizing that God is void of self-nature, not realizing that God, as the essence-of-all-things-transcending-all-things, as *Ein Sof,* guarantees that all things are also void of self-nature—or as we Muslims say, that there is 'no god but God', no other THING sitting there next to the One Thing that alone is, and, in Its aloneness, is no-thing. (I might be wrong about the Jewish Buddhists; many Jews may seek out eastern religions simply because the Judaism in which they were raised teaches 'how to be a Jew', but so often no longer really believes in God—except maybe as the Archetype of Jewishness.)

That's the *shunyata* side of it. From the *tathata* side of it, *personhood* runs through all things in the universe. Little particles reacting to each other in the same field are beginning to develop little eyes which later become the eyes of the dragonfly, the eyes of the frog, and the eyes of you and me, here in the 'human state hard to attain'—the only state which can make something called a Vow. As Olivier Clement recounts, in *The Roots of Christian Mysticism*, in the words of some Frenchman who had a vision of *tathata*: 'I could have given a name to every puddle in the road.' If personhood were not intrinsic to the voidness of things, then wisdom would not be inseparable (like it is) from compassion: only persons can extend compassion, and receive it. As Lew Welch said, 'WE'RE ALL THE SAME PERSON.' That's *tathata* as indistinguishable from *shunyata*; that's things being exactly what they are not *in spite of* the fact that they are nothing in themselves, but *because* they are nothing in themselves. When Lew said, 'I try to write from the poise of mind which lets me see that things *are* exactly what they *seem*,' he was saying that, if all things are mind-things, *what* they are (*shunyata*) is ultimately indistinguishable from how they *seem* (*tathata*); things are real *because* they are apparitions; in their very suchness they are unique and incomparable—they are better described as *beings* than as things. So the Void is also a Being; that No-one is also a Someone. 'Form is emptiness; emptiness is form'—which also means that if we somebodies are really Nobody, then that Nobody is also, in one sense, on one level, Somebody. You said that the Buddhists believe in leaving things entirely open, in not limiting reality to any particular Ultimate Form. But that kind of openness is what a Person really is. A Person is not my idea of him plus his idea of himself; that's merely the distortion and imprisonment of that Person by my ego, by his ego. A Person is the *form* of emptiness—completely open. The ego is a form that doesn't realize its own Emptiness. God is an Emptiness that com-passionately comes into Form in order to manifest Emptiness, which is close to what the Eastern Orthodox Fathers mean when they say 'God became man so that man might become God.' (We say 'Emptiness', but *shunyata* is only empty of specific self-determinations. Being completely open, being undimin-ished by any self-determinations, it is also completely full, oth-erwise it could not be called 'the Void eternally generative'.)

The (exoteric) Christians idolize the Personal God because they think He guarantees that this little me will always be the same little me, in heaven or in hell. The Buddhists turn up their aesthetic noses in refined distaste at such antics—as well they should! What they should not do is deny that Buddhism has its own rendition of the personal face of Absolute Truth, just like we Children of Abraham. The Dharmakaya is very much like the Godhead or that God the Father (especially in its Eastern Orthodox rendition) Whom 'none has seen at any time' (according to St. John the Evangelist); the Sambhogakaya is somewhat like 'God Almighty, Our Father in Heaven'; and the Nirmanakaya is precisely like Jesus, 'the Word made flesh', the human embodiment of the Dharma. And Lady Prajñaparamita, the 'field-aspect' of Wisdom, is very much like the Holy Spirit, or the Blessed Virgin (whom the Eastern Orthodox identify with *Sophia*—precisely 'Highest Perfect Wisdom'). Of course there are differences, but they are differences in dialect, not final intent; in word, not (deepest) meaning.

All forms arise in the One Mind. But it is the singular or multiple apparitional forms of the Dharma, and the form of my human teacher, that have the power to save—unless I'm next to enlightened already, in which case anything at all could do the same—for example, a flower held in an extended hand. But in terms of effective means, it is infinitely more likely to be the apparitional shape of God Almighty, not the apparitional shape of my glass of beer, which takes me beyond myself. God is the Wisdom/Means of the Void. While there is still (apparently) a little me, there is also a vast He, just as real as I am—realer, even, since the little me is the ego, while the vast He is the face of Absolute Truth as addressed to that ego. But when that He comes around and starts looking out through the eyes of the little me, then there is no me to see or be seen. And no He.

So He makes Himself useful. In Muslim terms, through Mercy. In Buddhist terms, through Compassion.

Sincerely,
Charles Upton

Christian & Muslim 'Trinitarianism'

A Reply to Philip Sherrard

Orthodox Christian writer Philip Sherrard has sometimes been identified as a member of the so-called 'Traditionalist' school, founded by René Guénon and brought to profound fruition by Frithjof Schuon. And yet in his last book, *Christianity: The Lineaments of a Sacred Tradition*, in the final chapter entitled 'The Logic of Metaphysics'—though certainly not in the greater part of the book—he seems on the verge of breaking with them over the fundamental doctrine of the Impersonal Absolute, which Guénon most often understands in terms of the Vedanta, but which is to be found in every sacred tradition, including Christianity. Certainly an idolatry of the Formless Absolute as a mental concept, often with the motive of avoiding both the efficacious warmth of devotion to God and all generosity and compassion in our relations with our fellow human beings, is a besetting sin of many attracted to the study of metaphysics. And the cold, mathematical precision of Guénon's writings reveals a personality which, while proper to his specific mission, should clearly not be taken as a model for most believers, or even most mystics, Muslim or otherwise. One wonders if Sherrard, in the face of the recent turbulence in the Traditionalist world, and under the shadow of his own impending death, may have mistaken the *sang froid* of a French metaphysical intellectual for the supposed heartlessness of the 'desert' of the Godhead, the Formless Absolute.

But such speculations as to Sherrard's motives are in a way unfair. His opposition to the idea of a Formless Absolute was

undoubtedly sincere, and based on specific metaphysical objections to Guénon's position. It is these objections alone that I will
essay to answer; as we Muslims say, 'may God keep his secret.'

Variety and Paradox
in Christian Theistic and Trinitarian Language

Metaphysical discourse must walk a fine line between a total
dedication to Truth and a realization that all metaphysical statements are 'operative' in a sense, ultimately justified only by the
effect they produce on our consciousness. A statement that is
'true' but does not *convey* Truth in no way serves the goal of metaphysics, which is to prepare the heart not to resist, through an
attachment to error, the direct intuition of Divine realities. This
is why the Buddha considered such questions as whether or not
human consciousness survives bodily death, or whether the
Tathagata does or does not exist after attaining Nirvana, as 'tending not to edification'. To the Buddhists, doctrine is an aspect of
upaya or operative method. If it breaks our attachment to erroneous ideas, it is true; if not, then not. On the other hand, the clear,
objective and orthodox expression of Divine Truth, even
though it is incapable of totally encompassing the Truth it
expresses, is one of the most powerful of *upayas*, which is why the
Buddhists, even while repudiating metaphysics (as the Theravadins and the Zen practitioners do) also produced metaphysical
systems of great subtlety and complexity (as, for example, that of
the Vajrayana school, or the Madhyamika). To idolize metaphysical formulations as if they were absolutely true, rather than
'quasi-absolutely' necessary within a given context, is to miss the
Truth by the positive road, which is literalism; to take such formulations as mere utilitarian tools for the manipulation of consciousness is to miss it by the negative road, which is nihilism.

Apophatic mysticism is the remedy for the first disease, cataphatic mysticism for the second—and this reciprocal relationship between metaphysical speech and metaphysical silence can
be applied to Christian doctrine as well. Thus when Christians

say that God is personal, they are indicating that everything we know as personhood is derived from Him, Who is nonetheless greater than any sense of personhood we can conceive of. Insofar as Eastern Orthodox theology denies the precedence of an impersonal Absolute over the persons of the Trinity—if it really does—this is to prevent Christians from forming a false image of an abstract, impersonal 'beyond-being' in the face of which God's personhood dwindles to 'mere' personality. But then the nearly universal testimony of the Christian mystics arises— including that pillar of eastern Orthodoxy, Dionysius the Areopagite, in his *Mystical Theology*—who tell us of a 'Godhead', a 'dazzling obscurity', a 'Divine desert' in which no forms can be discerned, including those of the Trinity, over which, in the mystical *extasis*, it is seen to take precedence, while in no way negating or subordinating this Trinity.

How can the Godhead take precedence over the Trinity without subordinating it? If we place logic above intellection, there is no way it can. If, however, we give intellection precedence over logic, then the problem simply disappears. If the doctrine of the absolute unity of the Divine Essence with its trinitarian embodiment protects the Christian sense of the Absolute from an abstract impersonalism, the testimony of the apophatic mystics to the transpersonal Godhead equally protects the sense of God's Personhood from a literalistic belief, leading to a sentimental personalism fundamentally indistinguishable from polytheism, that the Divine Persons can somehow be encompassed by human understanding—as well as guarding in its own way the notion of God's Essence against a lapse into the abstract and impersonal, since it is into this very Essence, an Essence which is both their original Source and their ultimate Reality, that the Persons are subsumed in the apophatic ecstasy.

The sense of the Trinity as a possible object of knowledge is essential to the purity and efficacy of Christian doctrine. And yet, in a truer sense, the Trinity cannot be an object of *our* knowledge, since it is precisely the necessary mode of God's knowledge of, and love of, Himself. In one sense the Son is the form of the Father's self-knowledge and self-love in eternity, while the Holy Spirit is the act of that self-knowledge and self-love. In another sense, the Son, from the Christian perspective, is Jesus Christ, the unique incarnation of God's eternal self-knowledge and self-love in space and time. But in a third sense—which cannot be separated from the other two without lapsing into heresy from a Christian standpoint—the Son is precisely the human being *in divinis*, as when St. Paul says 'it is not I who live, but Christ lives in me,' or Jesus answers the accusation of Pharisees that he claims divinity for himself by saying 'do not the scriptures teach that ye are all gods and sons of the Most High?' The Second Person of the Trinity would thus be analogous to the Sufi concept of *al-Insan al-Kamil*, the Perfect Man, which is both the archetype of humanity *in divinis* and site of manifestation of all the Most Beautiful Names, and the perfection of the human being who has realized his or her archetype. In this sense, the Trinity is precisely the relationship between man and God which pertains after man has attained *theosis*. If God became man that man might become God, then the Son is none other than the Christ in me. The Father is the God Whom, even though by Christ's power I have attained Godhood, I worship as greater than myself, as when Jesus said 'why do you call me good? None is good but the Father.' And the Holy Spirit is the essential Love and Knowledge Who unites the Christ in me with the Father Who is eternally above me; one might almost say that, after *theosis*, (and, virtually, even beforehand), the Son is myself as a son of God and co-heir with Christ; the Father, the Unseen God Whom I acknowledge as having precedence over me in terms of 'procession', though He and I are of one Essence; and the Holy Spirit, the very Unity of Father and Son. Yet 'it is not I who live, but Christ lives in me.' In my created humanity I am in no way God. Deification is not the perfection of myself as creature—to claim such perfection would be to rival God—but the realized perfection of the Divine within me, by virtue of the Second Person of the Trinity, the archetype of my Humanity, 'made in the image and likeness of God.' When Jesus said 'be ye perfect as your Heavenly Father is perfect,' he was not requiring that 'flesh' make itself equal to 'Spirit', but rather, through the

Son Who lives in the center and depth of the soul, that the perfection of the Father be realized, since 'I and the Father are one.'
(NOTE: The irreducible theological barrier between Christianity and Islam, even Sufic Islam, appears to be the *exclusive uniqueness* of Jesus as expression of the Human Archetype *in divinis*,
though the statement of Jesus that 'ye are all gods and sons of
the Most High' would seem to deny this literal exclusivity. In my
opinion, only Frithjof Schuon's doctrine of the transcendent
unity of religions can resolve this dichotomy—though not in
the world of form. God blesses each unique and spiritually-oper-
ative perspective He has established for our salvation—that is,
each religious revelation—with His Own Absoluteness, thus
rendering each of them, in Schuon's terminology, 'relatively
Absolute'. Each, from its own perspective, is the Absolute Itself.
These 'perspectives' are not subjective, however; they are not
based upon belief: they are as objective as seven different trails
up the slopes of Mt. Qaf, all headed for the One Summit.)

The fullness of the Godhead is present in all three Persons—
and yet there is still a precedence between them. If this were
not so, Eastern Orthodox theology would not insist so strongly
that the Holy Spirit proceeds not from the Father *and* the Son, as
in Catholic theology, but from the Father alone. The Father
takes precedence, as a hypostasis of the Formless Absolute, the
Divine Essence as It is in Itself, over both the Son and the Spirit.
And yet the fullness of this Divine Essence is equally present in
all three Persons. The Spirit is the infinite radiance of the Father;
consequently, in the Orthodox formulation of the Trinity, He
would be related to Guénon's (and Schuon's) Infinite, as the
Father is to Schuon's Absolute. And if the Spirit is Infinite Possibility, then the Son is Perfection (Schuon's third divine 'hypostasis' along with Absoluteness and Infinity), the perfect synthesis
of those Infinite Possibilities, and thus the perfect mirror, or
image, or *icon* of the unknowable Father, as when Jesus says both
'none have seen the Father at any time' and 'who has seen me has
seen the Father.' In terms of Divine Manifestation, the Father or
Formless Absolute must come 'first', (though first neither in
time nor in essence), the Holy Spirit second, as the radiance of

this Absolute in terms of Infinite Possibility, and the 'only-begotten' Son third, as the Father's perfect self-expression, 'without whom nothing that is made was made,' since all manifest existence is necessarily derived from the form of God's eternal self-knowledge. Nonetheless, in terms of redemption rather than creation, the Catholic formulation of the Trinity, in which the Holy Spirit proceeds from the Father *and* the Son, is also true, since even though the Holy Spirit is sent by the Father, this Spirit cannot be hypostasized as a distinct Person without the relationship *in divinis* between Father and Son, Divine Source and Divine Manifestation, any more than this Spirit can manifest in human life without the *relationship* between deified Man and the unseen Father—without God's grace and the human response to it—without prayer. In Christian terms, prayer is the communion of the Son within me with the Father above me; in a certain sense, prayer itself *is* the Holy Spirit. But this in no way makes the Spirit dependent for Its deployment upon the attitude of a mere creature, since the relationship between the unseen Father and Christ the Divine Man subsists in eternity, before the foundation of the world. In Catholic formulation, then, the Father is the Formless Absolute in the mode of polarity with Its own manifestation *in divinis*—this relationship being a distinct Person, the Holy Spirit—while in the Eastern Orthodox formulation, the Father is this same Absolute as It is in Itself, paradoxically beyond all relations while at the same time being the Source of all relations—the Source of all relations precisely *because* It is beyond all relations.

Some Specific Points in 'The Logic of Metaphysics' Answered

(1) In his chapter 'The Logic of Metaphysics', Philip Sherrard maintains that, according to Guénon, the Absolute is beyond logic yet can be 'typified' by logic. This logic is Aristotelian and 'exclusionary': A cannot be both B and not-B. Sherrard faults Guénon for applying exclusionary logic to the Absolute. But

then, (2) he does the same thing himself when he criticizes Guénon's formulation that 'Being is determined by nothing; it determines itself.' In trying to catch Guénon in a logical contradiction, he is himself applying exclusionary logic to the Absolute, thereby falling into the same error he attributes to Guénon. An undetermined which determines itself is a logical contradiction, true; but the fact is that logic does not apply on the level of the question 'how can a totally undetermined non-dual Absolute generate duality and determination?' Only paradoxical language can be applied to this 'mystery', which, as *maya*, will always escape a strictly logical definition.

(3) Sherrard understands that a purely negative description of the Absolute as non-determined (non-constrained) contains a subtle determination (constraint): the constraint of the Absolute that would seem to prevent it from generating determinations/constraints. Ibn al-'Arabi has dealt conclusively with this point, in his doctrine that 'the Absolute is not delimited by its own non-delimitation.' But then,

(4) Sherrard expresses this apparent contradiction by saying that the Absolute and the non-manifest divine Essence should not (as in Guénon) be identified, and does so using what he calls a 'radical' apophatic language, according to which neither positive nor negative language can be applied to the Absolute. But he makes the error of placing Being and Not-Being on the same ontological level as the two sides of a horizontal paradox, whereas the truth is that Non-Being is higher than Being, and the identity of Being and Not-Being is higher still. Thus he says 'If the Absolute is free from determination, it is also not free from determination,' whereas it is more accurate to say that 'If the Absolute is free from determination, it is also free from non-determination'; an Absolute which is 'not free from determination' is in no way Absolute. In Buddhist terms, the realization of Nirvana liberates us from Sangsara, while the realization of the identity of Sangsara and Nirvana liberates us from a reified and subtly conceptual Nirvana. Likewise Being is the truth which liberates us from contingency (religious faith saves us from the world); Not-Being, that which liberates us from an attachment

to Being (apophatic mysticism overcomes the subtle self-worship inherent in our worship of God on the plane of form); and the understanding of the Absolute as transcending both Being and Not-Being, that which liberates us from a one-sided mystical transcendentalism (the origin of heretical Gnosticism) which denies that the Absolute is capable of encompassing and manifesting Itself as the contingent and determined without losing its absoluteness (plenary esoterism overcomes the excesses and imbalances of apophatic mysticism). So Sherrard is right in saying that the Absolute transcends even Essence, if we take Essence as excluding all determination, but he is clearly wrong in denying that Essence is higher than determination, and asserting that it is simply the other side of the paradox 'Unmanifest Essence/Manifest Determinations'. To put it succinctly, the ultimate identity of Being and Not-Being can be realized only through Not-Being, not through Being, though subsequent to this realization it is certainly *manifest* through Being. And if we identify 'Not-Being' with 'Beyond-Being', as Guénon did, then we can't take it as simply the horizontal opposite term to 'Being', but must understand it as ultimately qualifiable neither by 'Being' nor by 'Non-Being', and thus as identical with Sherrard's 'Absolute'.

(5) Sherrard, somewhat puzzlingly, faults Guénon for being a 'pantheist' because his doctrine implicitly denies the reality of manifestation. But is not pantheism the doctrine according to which manifestation is mistaken for its Principle, and thus taken as a kind of a spurious absolute? One would have thought that accusations of Gnosticism or Manichaeism would have come more readily to Sherrard's mind—though Guénon is no more a Gnostic (in the sectarian, heretical sense) than he is a pantheist. Furthermore, when Sherrard places Being and Not-Being on the same ontological level as aspects of the one Absolute, without hierarchicalizing them, he is more of a pantheist than Guénon is.

(6) Like many today, Sherrard can only see a flat contradiction between the values of the particular and personal and the spectre of the Non-dual Absolute. He seems to believe, in

Chestertonian mode, that the personal is somehow horribly negated in this Absolute, as if the Godhead of the mystics were a desecration of all that is good and genuine in human life, a kind of total alienation. The Non-dual Absolute, however—*Nirguna Brahman*—is not so much impersonal as *Transpersonal*—otherwise the personal God could not be Its first intelligible manifestation. Likewise the personal God—*Saguna Brahman*—is not limited by His personhood, but 'open within' to the Transpersonal (as, in fact, all persons are, who otherwise would be nothing but limited and predictable sets of physical and psychic characteristics: caricatures, not characters). In the words of Ramana Maharshi, '*bhakti* is the love of God with form' (*Saguna Brahman*); '*jñana* is the love of God without form' (*Nirguna Brahman*)—and where there is an 'impersonality' which negates personhood rather than transcending it, there can obviously be no love. Love requires Person as its object; it equally requires the Transpersonal dimension, which is all that prevents the Person of God—or of one's human beloved—from being treated as a known and defined quantity, reified, and consequently imprisoned within the shell of one's ego, rendering the loved one drearily predictable, all-too-well known.

Trinitarianism in Islamic Metaphysics

According to the Islamic testimony of faith, 'there is no god but God.' Exoterically, this means there is only one Supreme Being; esoterically, this indicates that there is only one Being; God alone possesses the attribute of Being, along with all other attributes; you and I are nothing in ourselves—yet we appear, we act, we love, we suffer, God sends us prophets and Books to save us, and we have the power to submit to His Will—though not, paradoxically, unless He Himself wills it. So who are we? We are none other than Him (who else could we be, there being none else?), but only insofar as we attribute no Being to ourselves, and all to Him. (To attribute this Being to 'ourselves and all things' rather than to Him would be the error of pantheism.)

The responsibility to submit is ours; the command to submit, and thus the reality of submission, is His. If we were to possess being and essence of our own to place beside His Being and Essence (this being the sin of *shirk*, analogous to the Christian 'sin against the Holy Ghost'), then God would be a capricious tyrant, a personification of fate or literalistic predestination, since He would either submit or not submit *for* me, dooming me to Paradise or the Fire without action or obedience on my part. But since my Essence is none other than His, His Essence does not stand, will, or exist apart from or over against mine; He in no way limits my choice; I am free with His freedom to participate either in the eternal Mercy or the eternal Wrath of His Nature, according to the name of God I am from all eternity, a name both freely chosen (since there is no Being in me but God's Being, and God is free) and eternally destined (since God is Who He is, and cannot be otherwise).

From the point-of-view of exoteric Islam, trinitarianism is seen as tri-theism. This is obviously a misunderstanding of Christian doctrine, yet it is accurate to one of the ways Christian doctrine can degenerate or be misunderstood—like the view of the iconoclasts that icons are intrinsically idols. In essence they are not idols, otherwise any sensible or even intelligible 'sign' of God would be so too, including the Holy Qur'an; just as certainly, in common with all sensible or intelligible forms, they can be transformed into idols through ego-identification.

From the Muslim standpoint, to say 'Trinity transcends Unity' is an error, as I believe it also is from the Christian standpoint. 'The single nature of the Three is God' says Gregory Nazianzen. 'In regard to His oneness he is the Father. The others come from Him and return to Him without being confused with one another. They coexist with Him, without being separated in time, in purpose, or in power.' If the single nature of the Trinity is God, Who in regard to His oneness is identified with the Father, then Trinity cannot transcend Unity, but is rather the eternal expression of Unity.

Sufi metaphysics, as I understand it, would include the doctrine that trinity is the necessary manifestation of Unity according to the form of the act of cognition (mine, and God's too; it is not exclusively subjective, or exclusively cosmic), and thus that it is—as in Schuon's *maya-in-divinis*—necessarily virtual within Unity; to affirm the Unity of God is to *participate* in trinity. Certainly if 'the Unity of God's Essence' is seen as a mere abstract 'Godness' which must *logically* be possessed by the three Persons, since all three are God, then it no way transcends these Persons. They are concrete; it is a mere abstraction from that concreteness. But that's not what—or Who—the Unity of God's Essence is. By 'essence', here, I do not mean 'quiddity', the Islamic *mahiyya*, but the Absolute Essence, the Islamic *Dhat*. *Mahiyya* relates to the question 'what is it?' and must be answered by comparing the object in question to other objects: X is what it is because it is like A and not like B, etc. God certainly transcends *mahiyya*; God is not this or that object. *Dhat*, on the other hand, as Absolute Essence, is incomparable; It is incapable of being defined in terms other than Itself. When God said to Moses, 'I Am That I Am,' he was speaking out of that Incomparable Essence, identified with pure Being—which, we must remember, is in another sense superessential and Beyond Being.

Islam also has a doctrine of God as Being and Beyond Being. Being is *Allah*; Beyond Being, the Divine Essence, is *Dhat* or *Hu*. In one way, *Allah* is the Personal God Whom Schuon identifies with pure Being. Yet the word *Allah* literally means 'The Deity'— an impersonal name—while the designation for the 'impersonal' Essence or Beyond Being is *Hu*, which means 'He'—a personal name. The many names of God are names of the Essence; their totality and synthesis is *Allah*. In one sense they are qualities, attributes, and thus relatively abstract vis-à-vis the Essence; in another sense they are precisely *names* which, like all names, denote the essence of someone insofar as it transcends that someone's definable attributes. This is one of the ways the paradoxical *dance* (cf. the Christian *perichoresis*) of the Impersonal/Personal Transpersonal God is rendered in Sufi metaphysics.

The fundamental Muslim testimony of faith is 'there is no god but God'—and if I, through my essential nothingness, am His slave (*abd*), and thereby his fully-empowered representative in

this world (*khalifa*), and thus (finally) His symbol, whose arche-
type is the Perfect Man, *al-Insan al-Kamil*, God's first intelligible
act of Self-understanding from all eternity (according to Ibn al-
'Arabi, al-Jili and al-Iraqi), He who *is* the cosmos *in divinis*, just as
I, as Man, am the compendium of the cosmos in creation; and if
I (as Servant), Allah (as Lord, a Name which can only be applied
to Him if a Servant exists whose Lord He can be), and the Divine
Reality which unites us, share the one and only Being and
Essence, which is God's, then this is precisely the esoteric trinitarianism within Islam (or one of several versions of it), in which
the trinity is not, however, an intelligible and revealed Divine
Object, with myself as the subject viewing it, but is precisely
God's act of Self-understanding within me, and *as* me. From this
point of view, possible in terms of esoteric Islam, and perhaps
not fundamentally opposed to Christian trinitarianism, though
necessarily possessing a different nuance and probably lacking
true spiritual efficaciousness in a Christian context, the trinity is
not a the 'literal' nature of a God Who, in reality, is in every
sense other than me, being One without a second; rather, the
act of intellection—which in the inner sense is God's eternal act
of Self-understanding, inseparable from His Nature, and in the
outer sense both the emanation of the cosmos from God and its
reintegration in God—is itself trinitarian. The act of intellection
in divinis is the archetype of the Divine act of creation. According
to Ibn al-'Arabi, the Divine Essence eternally polarizes into Lord
and Servant, which in essence are none other than Reality. The
Essence of Lord, Servant and Reality, or of God, Cosmos and
Reality, is none other than God's Essence. Yet, in terms of form,
I remain God's servant, just as the cosmos remains God's creation; both servant and cosmos are totally contingent upon Him
alone. (See the *Fusus al-Hikam*, the chapter 'Salih'.)

So at the serious risk of mixing doctrines proper to different
traditions in the wrong way, I am compelled to declare that
Meister Eckhart's formulation 'The eye through which God sees
me and the eye through which I see Him are the same eye'—
Eckhart who truly was the 'Christian Ibn al-'Arabi' if I may so
express it—defines a trinitarianism which I, as a Muslim and

dervish who holds to the metaphysics of Ibn al-'Arabi, can fully accept:

'The eye through which God sees me' (the Father Whom 'none has seen at any time'; the Divine Essence in the sense of the Unwitnessed Witness, the Vedantic *atman)* 'and the eye through which I see God' (the Son; the site-of-manifestation of the Divine Essence; the primordial human nature, *al-fitrah,* as expressed in the *hadith* 'heaven and earth cannot contain Me, but the heart of my loving slave can contain Me', and in words of Jesus, 'who has seen me has seen the Father') 'are the same eye' (the Holy Spirit; the Universality of God; the Unity of Father and Son in the Absolute Reality.) In Ibn al-'Arabi's terms, 'I' am the Servant, 'God' is the Lord, our common 'Eye' is the Reality.

The Trinitarianism of Frithjof Schuon

In my opinion, all that Christian theology can (and should) nail down in terms of dogma is that the Deity is One God in Three Divine Persons. The dispute—which Bishop Kallistos Ware, for one, de-emphasizes—between the Catholic formulation where the Holy Spirit proceeds from the Father and the Son—the famous, thorny *filioque*—and the Orthodox formulation, where the Spirit proceeds from the Father alone, is simply one of perspective. As I indicated above, the Orthodox formulation apparently sees the Holy Spirit as the creative and saving radiance of the Father, which nonetheless comes to man through the Son, whereas the Catholic formulation sees the Holy Spirit more as the relationship between Father and Son, as in prayer.

The Traditionalist diagram can be helpful here. In general, the Traditionalists tend to speak schematically of the Father as the midpoint of the circle, the Holy Spirit as the radii, and the Son as the circumference. Insofar as the radii radiate from the Center, the Holy Spirit proceeds from the Father. Insofar as a 'radius' is precisely a straight line which unites Center and Circumference, the Holy Spirit proceeds from both Father and Son. The Father is the Source of the Spirit; but if there were no Son, there would be no recipient of that Spirit, which would never become intelligible, never be deployed as a distinct hypostasis; consequently the name 'Holy Spirit' could never be known.

The Church Fathers themselves speak of the Trinity from many different perspectives. The Father is the oneness of God, or else the engendering force within the Divine Nature. The Son is the creative ray of the Father, yet everything God does is done by the Spirit. The Son is the perfection of all things, yet the Holy Spirit is the one who brings all to perfection. What I see in this apparent mass of contradictions is the truth, which Schuon expresses so well, that metaphysics is suggestive, not dogmatic; it must speak from a multitude of perspectives, since by nature it is attempting to express, or give intuitions of, something which cannot and should not be made explicit. It can and should be expressed in terms of logic, insofar as is possible—and such expression is far more possible than is generally believed— but it cannot be trapped in logic. Theology is the necessary vessel of metaphysics; the metaphysical multiplicity of perspectives is no excuse for heterodoxy; it doesn't mean that you can say anything you want to; it is not subjective or impressionistic; some formulations are simply wrong. But if the Church Fathers can present many different versions of the Trinity in their attempt to render that mystery—versions which as theology would be contradictory, but which as metaphysics are precisely paradoxical—then perhaps the trinitarianism of Frithjof Schuon, based on the three hypostases the Absolute, the Infinite and the Perfect, and which is also situated on the plane of metaphysics, not that of theology, is no less orthodox.

The Absolute, the Infinite and the Perfect, for Schuon, are three and not three. Infinity is not Infinity in isolation, but precisely the Infinity of the Absolute, which is not other than Perfection. Neither is Absoluteness isolated in its own nature, being the essence of both Infinity and Perfection. Nor can Perfection be separated from either the Absolute or the Infinite, since it is the site of their manifestation and the Vessel of their Union. These three hypostases are expressed in impersonal terms, yet in Schuon's doctrine they also have a personal aspect.

Absoluteness, the principle of transcendence and thereby hierarchy, is the archetype of everything masculine. Infinity, the principle of immanence and thereby all-embracing and all-constituting Substance, is the archetype of everything feminine. And Perfection is the archetype of Divine Form, in which all manifestation, as well as manifestation's unseen Source, are gathered into perfect synthesis, *al-Insan-al-Kamil* or Perfect Man, the secret of God's eternal self-knowledge in the depths of His nature, which is the prototype *in Divinis* of the human form. Certainly this is not Christian theology any more than it is Muslim *kalam*, nor is it meant to replace them. Its specific function is to protect the providential and necessary formulations of theology from petrifying into idols, and also to point beyond them, directly to the Object of which they are the signposts here in this world. It is, precisely, metaphysics, which as Schuon points out is as much 'musical' as 'mathematical'. It is the dance of the human mind on the shores of the Inexpressible. It is the reflection of what the eternal Essence that expressed itself on earth as Philip Sherrard is gazing upon, God willing, even now. As death is swallowed up in victory, so hearsay, and even sight itself, are swallowed up in the taste, and the embrace.

Intrinsic
Crucifixion

Central to Christianity is the notion of Christ's Incarnation, Death and Resurrection as a special, unique act of Divine mercy and generosity. In Hinduism, the one other revealed religion that recognizes a number of Divine incarnations somewhat analogous to that of Christ (i.e., the avatars of Vishnu—Krishna, Rama etc.), the incarnation in question saves humanity by conquering evil forces and establishing and empowering a fresh version of the spiritual Path, but only Christ voluntarily sacrifices Himself, suffers and dies, to save the human race from the consequences of the Fall. In this lies the uniqueness of the Christic revelation, a uniqueness which must be understood in view of the fact that every revelation of God, as well as each Divine act, both throughout time and beyond time, is unique; as the Qur'an teaches, God never repeats Himself. (Purusha, the primordial Person in Hindu myth, is sacrificed and dismembered—like the Norse Ymir—to produce the 'ten-thousand things' of the manifest universe; Purusha, however, is not a saving avatar. But even though no avatar sacrifices himself to save the world in Hinduism, an analogy might be drawn between Christ and Purusha in view of the fact that Christ's Death and Resurrection is viewed as a 'second creation': 'Behold, I make all things new.')

But even though Christ's manifestation is both a free act of God and an incomparable instance of Divine mercy and condescension, it could never have happened, nor could it have been effective to save, if it were not in line with the nature of things; to follow the nature of things is to obey God, not first on the level of His commands and prohibitions directed to humanity, but on the level of His essential attributes, of which all things are reflections. Behind 'not my will but Thine be done' lies 'not my contingent nature, but Your absolute nature, is the Real.'

The central paradox of the Crucifixion is as follows: that since God is perfect, He could not suffer, given that suffering is imperfection and privation—but that, nonetheless, according to another perspective, God must necessarily suffer, both because mercy and compassion ('suffering with') are intrinsic to His nature as aspects of His perfection, and because he is immanent in all things, as the single Essence (though certainly not the formal quiddity) of all things. If God suffers, it is both because he loves all beings, and because He is the ultimate Witness, and therefore the ultimate Experiencer, of the sufferings of all beings. (To say that God 'loves all beings' is certainly not to deny His majesty, nor His role as the One Who judges beings with the inflexibility and rigor of the Truth. God loves all things in their original perfection as he conceived and created them, not in their rebellious divergence from that perfection; it is in fact this very primordial love which constitutes the rigor of His Justice in the face of that divergence.)

To declare exclusively that God is beyond suffering is to assert His Transcendence—which is most certainly true—but deny His Immanence. To declare exclusively that God, through Christ—or in His own nature, not only according to the Christic revelation, suffers the sufferings of all beings is to is to assert an 'Immanence' that compromises the Divine perfection by involving God—as the process theologians and others have done—in the contingencies of His creations, thus veiling the very Transcendence that gives Immanence its meaning, and making Him less than God. Only an intellective gnosis that understands Him as being both Transcendent and Immanent, without these two perspectives involving in any way a duality within the Divine nature, can grasp the intrinsic reality that Christ manifested by His suffering and death on the cross. (Likewise the 'bodhisattva vow' of Mahayana Buddhism, though exoterically considered as a voluntary sacrifice, on the part of the Buddha-to-be, of entry into Nirvana for the purpose of saving sentient beings—an entirely valid perspective on its own level—

may be understood esoterically as an instance of intrinsic rather than voluntary compassion. To the degree that the bodhisattva realizes the 'voidness' of his or her own self-nature—a realization that is analogous, though in a radically different mode, to Christ's voluntary self-sacrifice—he or she becomes the compassionate host to the illusory self-natures, and thus to the varied experiences, of all sentient beings, and to all the suffering that is part and parcel of such illusion; this is the meaning of the Mahayana doctrine that 'emptiness [shunyata] is inseparable from compassion [karuna]'.)

We know from our own experience that identification with and attachment to limited and perishable forms diminishes our capacity for suffering. To the contracted soul, immersed in the world—that is to say, in the ego—the slightest irritation is too much to bear—which is to say, too much to be impassively witnessed. One implication of God's Transcendence is that He is beyond all contraction, identification and attachment. And it is this very Transcendence that gives Him the capacity to suffer without being diminished, fragmented and lost in this 'experience'. It is precisely because God transcends all things that He possesses the power to 'suffer with' all things—a power that is inseparable from His Immanence. Thus His suffering is both infinitely greater than the suffering that any limited, created being is capable of, and also—in essence—no suffering. In this fact lies the mystery of compassion, of which Christ's death on the cross is the most concentrated instance of which we have knowledge. The self-sacrifice of Christ was, and is, an incomparable instance of God's compassion for the world; it is also perfectly in line with the nature of things, and thus inseparable from the intrinsic nature of God. It is unique, free, gratuitous and never-to-be-repeated; it is also inevitable: 'The Lamb' is 'slain from the foundation of the world'.

Experience vs. Being in the Mystical Life

A Meditation on Psychedelics

We have nothing to do even with the fantasies of the angels; what business is it of ours, then, to indulge in the fantasies of the satans? —Shams Tabrizi, on hashish

Enter houses by their doors. —Muhammad

The renewed talk in some circles about the use of psychedelics[1] to stimulate 'mystical experience' has led me to ask whether or not 'experience' is really central to the mystical life. It may be that the tendency to define mysticism in terms of a central 'ultimate' experience—as, for example, in the writings of Evelyn Underhill—has subtly falsified our conception of that life, divorcing it from the practice of virtue, the need to conform to tradition, and the understanding of metaphysical principles.

If earlier generations made virtue or confessional orthodoxy into idols—if, that is, the real nature of orthodoxy and the true goal of virtue were imperfectly understood—the last two or three generations have certainly done the same for 'experience'. The test of the validity of any religious commitment, or any human relationship for that matter, is now seen in terms of experience,

---

1. Although the term 'entheogens' is beginning to replace 'psychedelics' to denote the drugs in question, I believe that the earlier term is the more accurate. Drugs may have the power to 'expand (and thus attenuate) the psyche'; they do not have the power to 'generate God within'.

no longer in terms of honor or duty or strength of character. A marriage based on duty rather than affection is obviously far from ideal, but so is one based on the sort of affection which neither develops out of nor flowers into any sense of duty whatsoever. Without affection it is hard to be dutiful, but with no sense of duty or responsibility affection is ultimately reduced to a shared narcissism.

I believe that the greatest danger of psychedelics—greater than their undeniable potential for damaging the mind or opening the soul to demonic activity—is their power, in certain instances, to produce 'valid' mystical experiences: true insights into the nature of God and our relationship to Him. To some, this may seem like an absurd conclusion: Is not the direct experience of God, the 'beatific vision', the crown of the spiritual life? And if such a vision is admitted to be valid, then how could it be dangerous? How could it be wrong? If God is the Sovereign Good, how could witnessing that Good, by any means available, be anything but the greatest good life can offer?

I will answer this objection with an assertion: *The essence of the mature spiritual life is to love God's will above His gifts.* One might catch a glimpse through a window of a beautiful women disrobing, but if She has not granted us that glimpse then we are violating her dignity, not to mention breaking the law. I believe that the same principle *strictly applies* to the mystical life.

I will return to this point in a moment. For now, I wish to point out that the quasi-deification of experience is inseparable from the idolatrous worship of the human subjectivity. Duty is something we owe to someone or something beyond us, which is why fulfillment of duty is always potentially an approach to self-transcendence. Experience, however, is not something we 'give', but something we 'get'.

In terms of Sufi doctrine, 'stations' are acquired through the fulfillment of spiritual duties, which is another way of saying that we develop them through the actualization of virtues. Courage, humility, patience, zeal, contentment with God's will are spiritual stations; they are acquired through labor. Spiritual 'states', however, are not acquired—they are given. Since no amount or any quality of human action can 'add up' to a theophany of Absolute Truth—though such action can, up to a point, remove barriers to our receptivity to that Truth—such theophanies, such instances of mystical experience, are necessarily free gifts of God. This means that the attempt to 'get' these experiences, to have them for our own, to savor them for our personal pleasure and enlightenment, is tantamount to robbery. And to rob God is not to come closer to Him, but rather to become estranged from Him—perhaps for all eternity.

The mystical life has first to do with Being, and only secondarily with experience. Our duty is to come into a real, viable, relationship with Absolute Truth, a relationship which holds true *whether or not we presently experience it.* As Beat Generation poet Lew Welch wrote, 'I seek union with what goes on whether I look at it or not'—not with the subjective experience of the thing, but with the very thing itself. In the words of St. Paul, *faith* is 'the presence of things hoped for, the evidence of thing not seen.' In the mystical life, experience is the product of faith, not faith of experience.

There is no mystical life, of course, without experience—because God is generous, because the Absolute Truth, by Its Infinite Self-manifesting Radiance, must communicate Itself. We ourselves, in fact, are actual instances of that communication. Experience, however, is inseparable from the subjectivity whereby it is experienced—that is, from our ego—while the entire *raison d'être* of the mystical life is to transcend this ego. Thus the Sufis hold that specific spiritual states are given us by God not as permanent goods in themselves, but rather in order to deconstruct or annihilate specific aspects of our ego, which is why the fully-realized Sufi is described as being 'beyond states and stations'.

Mystical experience which is actively sought for itself alone, not to please the One who sends it—and He may or may not choose to send it at any given moment—is experience sought by the ego. Consequently it can in no way deconstruct that ego; it can only inflate it—*even if the initial experience is one of ego-transcendence.* Therefore it is entirely possible that mystical experience such as that produced by psychedelics may actually *drive us away* from the Reality we have caught a glimpse of, *to the degree that the glimpse we have caught is actually valid.* It's as if the beautiful woman we have seen disrobing realizes that she's being spied upon and calls the police. The net result of such an encounter is obviously not to bring us closer to her. It may turn us into stalkers, but never into lovers. (This is not to say that every artificially-produced glimpse if higher realities must always estrange us from them, only that every such glimpse must eventually be paid for.)

There is, of course, the level of reality where Being and Knowing are revealed as One, the level of the Transcendent Intellect or Absolute Self. But in order for Being and Knowing to unite, 'experience' and the 'experiencer'—which are all that keep Being and Knowing (apparently) apart in the first place—must be annihilated. God does not 'have experiences': He Knows, and Is. In the words of the Prophet Muhammad, peace and blessings be upon him, 'pray to God as if you saw Him—because even if you don't see Him, He sees you.'

.... when the king came in to see the guests, he saw there a man which had not on a wedding garment:

And he sayeth to him, Friend, how camest thou hither not having a wedding garment? And he was speechless.

Then said the king to the servants, bind him hand and foot, and take him away, and cast him into the outer darkness; there shall be weeping and gnashing of teeth.

—Matt. 22:11–13

Morality and Gnosis

The letter killeth; the Spirit giveth life

I

In some self-styled 'esoteric' circles, it is common to hear phrases like: 'The sage is beyond good and evil.' Often this is taken to mean that, by virtue of his wisdom and/or spiritual power, he has earned the privilege of doing anything he wants. Morality is for the unenlightened; the Sufi shaykh or powerful shaman or 'Crazy Wisdom Guru', in St. Paul's words, is no longer 'under the curse of the law.' This *antinomian* tendency may have been given a new lease on life in the west by people like Friedrich Nietzsche or Aleister Crowley, but in reality it is ancient; when the idea that sufficient magical potency gave one the *power* to break taboo with impunity was replaced by the idea that spiritual exaltation gave one the *right* to break sacred law, antinomianism was born. Certain heterodox Gnostic sects within both Judaism and Christianity have been antinomians, believing that esoteric knowledge absolves the elect from conventional morality; the famous Sufi Mansur al-Hallaj *apparently* violated the Muslim *shari'ah* by such statements as 'I am the Truth' (i.e., 'I am God') and was executed for them; and Jesus himself outraged the sensibilities of the Jews of his time by shocking acts such as driving the money-changers out of the Temple, and by declarations like 'before Abraham came to be, I am'—the last two words being the English translation of the first part of the Name of God in Hebrew, *Yah*.

It should be fairly obvious that the antinomianism of the magician Aliester Crowley is poles apart from the apparent antinomianism of Jesus and Paul, and in no way to be compared

with the ecstatic excesses of al-Hallaj. Among Crowley's dicta
were: 'Do what thou wilt shall be the whole of the Law' and 'Love
is the law, love under will.' His error, (intrinsic to the sin of lust,
among many others) was to believe that love can be ruled by self-
will. Paul's teaching, on the other hand, and Christ's, is that the
only way to overcome self-will is to place it under the rule of love.
According to Crowley, to be beyond good and evil is to reserve
the 'right' to break any law. According to Paul, only someone
ruled by love can perfectly fulfill the law, and thus escape its curse.

But if the sage or saint or master or avatar is indeed 'beyond
good and evil', what does this mean? Nowadays, whether or not
we admit it or are fully conscious of it, this phrase usually suggests to us a *right to sin*, a license to do evil. What else could it
mean to no longer be bound by the law, except that we now
possess both the power and the opportunity—and implicitly,
also the desire—to violate the law to our heart's content? In line
with this usually unstated belief, and to Satan's great delight, the
notions of *enlightened sage* and *sociopath* have begun to become
confused in the collective mind, a confusion that seems to be
validated anew whenever a religious figure is convicted of a
crime, or a criminal madman invokes a 'spiritual' mission or the
commandments of a divine 'voice' or to justify his bestial acts.

To be beyond good and evil, however, to be no longer under
the curse of the law, is not simply a license to kill. Rather, it is to
be bound by essentially *gnostic* norms rather than behavioral
ones. According to the well-known Sufi proverb, 'what is paradise to the believer is only a prison to the gnostic.' This is not,
however, because the gnostic is free of all constraints to his self-
will, but because he is bound by a higher and more rigorous law
than any that can be expressed in terms of explicit behavioral
rules. The more rigorous the law we obey—if that law is truly an
expression of God's Reality, and of His sovereign will through
which that Reality is made manifest—the greater is the degree
of freedom conferred by that very obedience. The believer is
commanded to avoid, repent of, and make restitution for *sins of
action*; the gnostic must do the same in relation to *sins of attention*.
When Jesus said, 'whoever looks at a woman with lust has

already committed adultery with her in his heart,' he was giving a concrete and easily understandable example a sin of attention, and thereby positing a higher, gnostic morality that transcends behavioral rules. To transcend the morality of action, however, is not to overturn that morality; as Jesus said, 'I come not to destroy the law but to fulfill it.' Nonetheless, those who are bound to a higher law than the external and behavioral will often appear to those whose level of understanding stops at outer, visible actions to be violating the sacred norms of religion; so Jesus appeared to the Scribes and the Pharisees when he performed miracles of healing on the Sabbath. In the Qur'an, in the *surah* of The Cave, the story is told of an encounter between Moses and a mysterious servant of God whose acts are shocking and incomprehensible, a figure the Sufis identify with their hidden and immortal patron, Khidr. Moses asks this person to be his guide, but the sage predicts that the prophet will not be able to abide his actions. But Moses insists, and so the sage takes him on. In their travels the pair come to the seashore, where the mysterious guide vandalizes a fishing boat; Moses objects. Next they encounter an apparently harmless young man, whom the sage immediately kills. Moses objects again. Finally they come to a town where they are spurned and denied hospitality, and the sage responds by repairing one of the walls within that town. Moses objects a third time, and the sage tells him that they must now part company; but before they do, he explains the purpose behind his apparently aimless and destructive acts. The fishing boat was vandalized to make it of no interest to a piratical king who was in the habit of stealing all the boats he encountered. The young man was killed in order to stop him from harming his parents, and so that God could give them a new son, pious and dutiful. And the wall was repaired so as to prevent a great treasure buried beneath it from being stolen before it came into the hands of two orphans, its rightful owners. Thus the actions of this mysterious master whom Moses could not abide are a shown as examples not of an antinomian immorality, but of a *gnostic morality*. They were based on *knowledge*; anyone who knew what he knew, God willing and permitting, would presumably have acted as he did. Clearly he did not act in obedience to his passions, or according to false or incomplete views of the situations he encountered; he acted on the basis of a higher Knowledge that was also a higher Mercy.

The major behavioral vices—anger, greed, lust etc.—remain vices in terms of gnostic morality because attachment to such desires *destroys unity of attention.* We are commanded to remember God, to love the Lord our God with all our heart, soul, mind and strength, to 'pray without ceasing'. We are called upon to make the living presence of God a constant reality to us, and to see all Being as a Unity in light of it. But if our attention is abducted by anger, geed and lust, whether or not we express these inner vices of attention in terms of outward sins of action, then we have violated that Divine command. Such vices are universally recognized as evil because their outward expression so obviously has a destructive effect on human society. But there are other sins of attention which destroy our sense of God's presence and our vision of the Unity of Being just as effectively as anger, greed and lust—and many of them, because their destructive effects on human society are not so obvious, are not even recognized as sins. Among these sins of attention are *fear* and *sadness, dullness* and *moroseness* (when we indulge in them as methods of denying and fleeing from the Truth and Love of God), as well as *cynicism,* self-indulgent *fantasies* of all types, *levity, giddiness* and *scatteredness.* All of these fracture the Unity of Being just as effectively as attachment to thoughts of lust or anger. And if we will take a moment to consider their ultimate effects, we may realize that they are just as destructive to human society in the long run as what are usually considered to be the major sins. When we consider the large percentage of our children, due to diagnoses of Attention Deficit Disorder and similar 'syndromes', must be drugged in order to keep them in school, the dire effects of mental scatteredness should become immediately apparent. To process too much information (as we are increasingly being forced to do) is just as much a mental or emotional disease, just as much a sin of attention, as drinking too much alcohol or eating too much food.

Certain uses of the mind may even have positive effects on society—or at least effects that initially seem positive—and still be sins of attention. A scientist or political activist who one-pointedly, but also obsessively and narrow-mindedly, pursues the solution to a scientific or social problem, may do much good. But he or she is also creating imbalances in his or her psyche, imbalances whose effects on society at large must eventually come due. Even the philosophical or contemplative pursuit of spiritual Truth may, under certain circumstances, become a sin of attention. In the words of Frithjof Schuon, 'mental passion pursuing intellectual intuition is like a wind that blows out the light of a candle.'

So we can see that gnostic morality, far from representing a relaxation of moral strictures earned by the great sage as a reward for his spiritual attainments, is actually a subtler, more rigorous and more all-embracing burden of moral duty than any mere behavioral moralist could ever imagine. It is a burden so great that, in the final analysis, only God can bear it.

II

To get a clearer idea of what sins of attention actually are, and how they relate to sins of behavior, let us consider them in terms of what, in the western church, have come to be called the Seven Deadly Sins: pride, avarice, lust, envy, gluttony, anger and sloth.

Pride is the luciferian counterfeit of union with God, Liberation, Enlightenment. To be proud of oneself, for whatever reason and in whatever context, is to claim to be God in one's separate selfhood, whether or not one is conscious of making this claim. The one united with God is not, however, united with Him in the sense of two real entities which have now become one, but rather is liberated from the belief in himself as an entity who intrinsically possesses Being, in the full knowledge that the only One to Whom Being can be attributed is God. Thus the true opposite of pride is not humility, but self-respect. To respect ourselves is to respect the God within us, Who witnesses us as we are—and the only way to perfect this witnessing is to realize the Witness as the Self, and the witnessed as nothing in itself other than this very Self, Who is God as 'I', not myself as 'me'. Pride is the origin of all the other sins, because without self-worship, the worship of anything other than God is impossible. In terms of behavior, pride licenses us to commit every other offense; in terms of attention, the essential act of pride is to place our attention on our own existence apart from God—that being the sin of Lucifer, the root of all the others. Pride, as self-worship, is the basis of all idolatry; whatever idol we may serve, whatever object we may worship in place of God, that thing only becomes an idol by *identification*, by our own act of associating it with our ego: every idol is a mask of 'me'. And here the self-contradiction hidden within the sin of pride becomes apparent: to worship oneself as God is to worship *another*—but God can never be 'another' to Himself, because God is One.

Avarice is the compulsion to solidify one's separation from God by identifying with one's possessions, to so completely surround and entrench and fortify oneself with material and psychological goods that one begins to unconsciously believe that one will never die, never be forced to let go of anything one has ever identified with. In behavioral terms it encompasses all forms of greed. In terms of attention, it is the compulsion to studiously avoid any true knowledge of oneself by treating oneself as a possession, an object; as such it is the satanic counterfeit of *jñana*, or contemplative objectivity. Instead of knowing oneself as objective to the Absolute Witness within us, the *atman*, the Eye of the Heart, we treat ourselves as quasi-material objects held fast by a reclusive miser who never sees or understands himself, only his precious gold. Avarice is the luciferian counterfeit of God's will and power to preserve the existence of His creation.

Lust is the act of denying the personhood of another (or oneself) for purposes of pleasure. As avarice transforms one into a soulless object, so lust transforms the other. Lust is a denial of the love of God. God loves us because He is totally aware of us, or rather aware of us in our totality. He sees us as He made us, in terms of our original and perfect conception, not as the partial and divided things we have made of ourselves (though He is certainly not unaware of these distortions). According to the *hadith* of Muhammad, peace and blessings be upon him, 'Pray to God as if you saw Him, because even if you don't see him, He sees you.' God can see us but we cannot see Him, at least in His totality; the sin of lust, however, falsely inverts this relationship. The lustful one either treats his 'beloved' as if he or she were not a conscious perceiver at all, only a kind of sex toy, or acts as if she (the object) had only an imperfect knowledge of him (the subject), while the subject's knowledge of her (the object) is perfect, resulting in a longing on the part of the object of lust for fuller knowledge of the lustful one—a knowledge that is forever denied her, producing in her a longing that the lustful one takes as a form of worship. In other words, the lustful one, in claiming to be the Knower but not the known, puts himself in the place of God. Alternatively, as we approach the 'masochistic' end of the spectrum, the one ridden by lust may desire to be encompassed by the greater reality and consciousness of his or her object, thus falsely worshipping the other instead of God. But in both cases, the reality of full, sincere and conscious human relationship is contradicted by the denial of the reality of either the other or oneself. As a sin-of-attention, therefore, lust involves an imbalance in attention between subject and object, a partial unconsciousness of one side or the other of the subject/object dichotomy, thus making their ultimate re-union, and thus the transcendence of the pairs-of-opposites, impossible. This sort of imbalance-of-attention is analyzed in another way by Martin Buber, in *I and Thou*, as either a flight from the subjective self into the outer world, or a flight from the outer world into the subjective self—two forms of alienation that make any encounter between a genuine 'I' and an authentic 'Thou' impossible. Lust is the luciferian counterfeit of God's love for and delight in His creation, which He knows as of one Essence with Himself.

Though *envy* can take the form, on the behavioral level, of the

intent to commit theft, it is perhaps the most mental of the
seven deadly sins; it can be defined as the obsessive sense that
what belongs to another should really belong to oneself (envy
proper), or that what was once one's own, or is some way *intrinsically* one's own, now unfairly belongs to another (jealousy). In
terms of the sin of lust, the need for union coupled with the
impossibility of it necessarily results in sexual envy. But we may
also be envious of the material, the psychological, even the spiritual possessions of another; in spiritual terms, to pry into the
secrets of God instead of waiting on God's good pleasure to
reveal or conceal them, especially when we feel that God has
unfairly gifted another with the knowledge we crave, is also a
form of envy. The other six deadly sins are usually defined in
terms of sinful acts; only envy is more or less clearly defined as a
sin of attention, which is sinful even if it never leads to theft, or
adultery, or detraction, or spying, or malicious gossip. Envy is
mental avarice, an avarice which blots the desiring self out in
the face of the desired object; it might be defined as self-loath-
ing objectified. The good we enviously desire is an idol to us,
since we worship it with our attention; but we also despise that
object (as we despise ourselves in desiring it), since we see it as
having no intrinsic value in the hands of another. Its only 'value'
comes from its perpetually incomplete identification with our
own unsatisfied, greedy little selves—and this (paradoxically)
indicates that the opposite evaluation is also at work, the feeling
that the desired object (another person's spouse, for example)
only has value because it is possessed by another; as soon as it
passes into our own hands it falls under the power of our self-
loathing, and turns to ashes. As a sin of attention on the psychic
level, envy often includes a 'paranoid' tendency to obsessively
analyze another's motives while remaining blind to one's own;
we spy upon and plot against the object of our envy, while
falsely seeing him or her as plotting against us. On the spiritual
level, the sin of attention inherent in envy is the tendency to
ignore both one's own intrinsic poverty and God's essential generosity, or to transform the knowledge of one's poverty into a
false sense of grievance, and God's generosity into a false sense

that He, or rather the world we worship in His place, is stingy and possessive; ultimately the envious one will see anyone's good fortune as his or her own loss. If we were to place our full attention upon God we would understand both His generosity and our poverty, and know our poverty as our greatest conceivable good fortune in the face of His generosity. Envy, in the sense that it claims the right to know every secret, is the luciferian counterfeit of God's omniscience.

Gluttony is the futile attempt to widen the area of one's own selfhood by devouring the world. Where avarice wants to possess and hold on, gluttony wants to incorporate. Gluttony is imperialistic, like a young king ruling an expansionist state; avarice is like an old king who wants to hold on to what he has already acquired, and to acquire more only to maintain his position, not to build it. Gluttony resembles an inverted generosity. Instead of giving out of our abundance, recognizing it as ultimately as God's, we revel in it, and end by robbing God of what He would have given freely, taking Him for granted instead of accepting His generosity with humility and gratitude. Gluttony is complacency, as in the phrase 'comfort food'; as a sin of attention, it is based on a denial of our intrinsic need, as creatures, for God's abundance, as well as on an unconscious *identification with God*, the One Reality Who is the nourishment of all, and to Whom all is nourishment. It is also a false identification of the world *with* God and His generosity; to the gluttonous individual, whether he be a glutton for food or a glutton for attention, 'the world is his oyster.' Gluttony is the luciferian and *inverted* counterfeit of God's creative power.

To fall into the sin of *anger* is to claim the divine right to determine what has the right to exist and what does not. As pride is a luciferian counterfeit of the Absolute, so anger is a counterfeit of God's omnipotence, which has absolute rights over all things, and both the right and the power to annihilate anything that believes it can oppose the Divine Will. As a sin of attention, anger manifests as an obsession with the existence of something, whether thing, person or situation, that it feels has no right to exist. Anger demands that any obstacle to the triumph of its will be annihilated *right now*; the self-contradiction inherent in this sin is that by its obsessive attention it is continually giving existence (as obstacle, not as essence), to the very thing that it wills not to exist; this is the frustration inherent in rage. In its most intense form, anger counterfeits the Divine Transcendence, in the face of which nothing 'other' than God has the right to exist.

The darkest of all sins is *sloth*. Pride is an attempt to overcome the subject/object split by seeing oneself as Absolute and Infinite; avarice is the attempt to overcome it by identifying oneself with and possessing the world; lust is the attempt to bridge the gap by possessing or being possessed by a desired object; envy is the attempt to overcome the subject-object split by identifying oneself with the possessions of another and losing oneself within that identification; gluttony is the attempt to overcome it by incorporating the world so as to transform the world into the self; anger is the attempt to overcome it by ruthlessly destroying all things that seem to come between the self and its world, one by one. But *sloth* is the attempt to overcome the subject-object split by sinking below it; it is an act of inverted Transcendence. Sloth is a seamless union of complacency and despair: complacency, because it is an acting-out of the proposition 'nothing need be done'; despair, because hidden beneath 'nothing need be done' is the opposite proposition, namely 'nothing *can* be done'. Avarice wishes to possess matter; gluttony wishes to incorporate matter; sloth wishes to *become* matter. If I am matter then I am safe, because matter/energy can never be created or destroyed, only changed in form. If I am matter then I am immortal, because matter is dead, and what is already dead can never die. As a sin of attention, sloth is based on a total and willful ignorance of spiritual reality (resulting in an ignorance of psychological and material realities as well), an ignorance that involves both the repression of the subject, since if there is no Spirit there is no consciousness of self, and the repression of the object, since if there is no consciousness in the subject, then no object appears. The other six deadly sins are based on partiality and perversion of attention; sloth is the complacent, despairing, and ultimately futile attempt to destroy attention totally. As such it is the luciferian counterfeit of the *Cloud of Unknowing*, the unknowable Essence of God.

The seven deadly sins, conceived of as sins of attention, can only be definitively overcome by gnosis, by the clear and objective understanding of the true nature of things. Behavioral morality serves this gnosis by restraining the passional actions that obscure it. Nevertheless, such morality can only be perfected by the full realization of gnosis itself, both because of the gnostic understanding that the ultimate motivations behind the passions are none other than erroneous beliefs about the nature of things, and because only in the established vision of plenary Truth can the struggling partiality of an existence defined primarily in terms of the apparently independent actions of its various creatures be transcended, and ultimately laid to rest.

III

In the myth of the Fall from the book of Genesis, to eat of the Tree of the Knowledge of Good and Evil is to fall from the level of a unified, *cardiac* consciousness that sees Truth directly, to that of a divided, *cerebral* consciousness based on discursive reasoning. Once sin divides our vision of reality in two, it becomes our duty to choose the better half; the fall into sin is also necessarily the fall into morality. And yet the effects of the Fall cannot ultimately be redressed unless we transcend the world of morality and return to Unity—a transcendence which, however, cannot be accomplished except through morality itself, at least initially: to choose the better half of a dichotomy is to choose what, in the world of oppositions and divisions, is nonetheless closer to Unity. Yet where *any* level of choice is still in force, the 'choiceless awareness' of metaphysical Unity cannot be attained; and the only way beyond the partiality of choice is perfect *islam*, perfect submission to the will of God. Morality is necessarily involved with choice, therefore it behooves us to choose wisely. Gnosis, however, is not choice but Vision; to witness Truth is to realize Unity, a Unity that embraces all conceivable dichotomies but is not determined by them. The relationship between morality and gnosis is expressed in the Qur'anic story related above, where Moses symbolizes law and morality, and Khidr, gnosis. It is true that Khidr vastly transcends Moses' conception of things; it is also true that if you want to meet Khidr, first you have to be Moses.

Among the dichotomies encompassed by the Unity of God is the one called 'good versus evil'. This, however, does not mean that God is somehow half good and half evil—as Carl Jung, for one, speculated in his *Answer to Job*. For the inconceivable reality of God to appear in relative, dimensional existence, dichotomies are necessary, given that polarity is the principle of all manifestation; without some kind of 'figure/ground' relationship, nothing at all would appear. And there is nothing that can highlight the immense and intrinsic goodness of God like the dark appearance of evil. To be 'beyond good and evil', therefore, is not to be half good and half evil, or to enter some nihilistic twilight zone where the appearance of good or evil is a matter of indifference. It is, rather, to realize the Sovereign Good. Partial good is always opposed to evil, and therefore involved with it, but no shadow of evil remains within the Absolute Good, because the Absolute has no opposite. The only way to be done with evil is to be done with the opposition of good *versus* evil, along with all the other pairs-of-opposites. Morality is the *Torah* of the Tree of the Knowledge of Good and Evil; gnosis of Absolute Reality is the *Torah* of the Tree of Life.

Gurdjieff, Ramana Maharshi, and the Name of God

I

Two of the most important practices in the Gurdjieff school, also called 'the Fourth Way', are, 1) 'self-remembering', and, 2) to distinguish and disentangle from one another the various 'centers' that compose the human psycho-physical being, so that each runs on its own proper energy.

Self-remembering is the attempt to remain conscious of oneself as distinct from, and in relation to, one's environment. The work is to keep what is perceived as self in a state of polar tension with what is perceived as other-than-self. This state of vigilance is distinct from any premature attempt to transcend the ego by identifying the self with the world or the world with the self—a double ploy which Martin Buber, in *I and Thou*, characterizes as a two-pronged delusional response to existential anxiety. If the ego is scattered in identification with the world of the senses, with social dynamics, with history—or, conversely, if the outer world is experienced only as a projection of the psychic subjectivity of the individual—then attention wavers; vigilance is destroyed; we fall asleep. Only the ongoing, lived encounter with the Other, the Divine or human 'Thou'—or at least a real world out there, different from us, which requires vigilance because it is dangerous and unpredictable—can keep us awake. (Many of Gurdjieff's shock tactics, and his work to identify and break people's unconscious habits and belief-systems, were undoubtedly attempts to command, at least temporarily, this particular brand of vigilance.)

But simply to will to remember ourselves is not enough; we can't keep it up. We discover that our will-to-remember is unable to tap the level of energy it needs to function in an ongoing way. We keep falling asleep.

According to Gurdjieff, the potential energy for wakefulness is bound up in the psycho-physical centers. Due to the inharmonious development of these centers, at least in civilized society, they often become confused with one another. When unnecessarily wired together like this they steal one another's energy instead of each center drawing on its own reserve. The centers are, let us say, the moving center; the sexual center; the feeling center; and the thinking center. If we think with our feelings, make love with our 'athletic' ability, move our body using mental will, etc., etc., then we waste the energy we would need to maintain a continuous state of self-remembering.

II

The spiritual method of Ramana Maharshi is 'self-enquiry'. On one level, this is simply the attempt to realize that 'I am not the body; I am not the breath; I am not the feelings; I am not thoughts; I am not the mind; I am not the I-sense', leading to the positive realization that 'I am Brahman; I am the Self'. However, as the Maharshi points out, the mind, though it may be able to dis-identify with what is other than itself, cannot dis-identify with itself: to say 'I am not the mind' *with* the mind is, simply, to continue to identify with the mind, while hypocritically denying this identification. 'I am not thought, I am not the mind, I am not the I-sense' remain nothing but thoughts, nothing but the action of the mind; they can in no way lead to the dissolution of the I-sense. In view of this limitation, the Maharshi recommended, as his essential method, the simple act of witnessing the arising and subsiding of the I-sense. According to Shankaracharya, the ego is a sign of the Self; the feeling that we are real selves is a sign of the Self within us. This, through it is often veiled by identification with the world of sense-objects and/or the world of thoughts, is nonetheless our most constant, obvious, and universal experience. But as Ramana Maharshi pointed out, if we attempt to isolate the core of this sense of being 'I', the I-sense dissolves. It was never there as a separate object or level of reality; it was nothing but the act of identifying the Self, the *atman*, with the body, or the breath, or the feelings, or the thoughts, or the mind. The rising and dissipation of the I-sense can only be witnessed by the Absolute Witness, the Self; the act of witnessing the I-sense is precisely what dissolves it, and invokes the reality of the Self. To affirm 'I am Brahman', is to posit a subtle duality; in the words of the Sufi Mansur al-Hallaj, 'Whoever declares that God is One thereby sets up another beside Him.' The practice is not to say 'I am Brahman'; the practice is, through the witnessing of the rise and fall of the I-sense, to *be* Brahman.

III

The Invocation of the Name of God—the Sufi *dhikr* or *zekr* ('remembering'), the Hesychast *mnimi Theou*, Hindu *japam*—is the essentially the practice of remembering God and forgetting oneself. (The Sufi form of this Invocation happens to be my own practice.) Yet Gurdjieff follower Jacob Needleman once assured me that 'what we call self-remembering, you call *zekr*.' Assuming he was right, what could he have meant?

To 'remember oneself' seems to require a splitting of the self into two parts: the rememberer and the remembered. But if the rememberer is part of myself, how can I remember it? I can look at my hands, I can look at my knees, but I can't look at my eyes. This leads me to suspect that the Gurdjieff method of self-remembering is meant to reveal its own inadequacy at one point, and by so doing invoke a higher sort of remembering. As Allah says in the Qur'an, referring to this transition: *Remember Me and I will remember you* (Q. 2:152). Self-remembering gives way to experiencing oneself as remembered or witnessed by God, which is the same thing as the recognition that we are perpetually being created by Him, instant by instant—and in ontological terms, if not according to the way our individual consciousness develops, God's remembering of us comes first. In the words of the Qur'an, *It is We who have sent down the Remembrance* (Q. 15:9).

The work of the Gurdjieff school, in teaching each psycho-physical center to operate on its own particular energy-source or frequency, is an attempt to free up more psychic energy for the act of attention, and also to make the elements or levels of the soul more capable of being witnessed as they are because they are no longer chaotically entangled with each other. (In reality, these are two different ways of saying the same thing, the first being *shakti* to the second. According to tantric doctrine, *energy empowers attention; attention liberates energy*.) In other words, I can only realize 'I am not the body' if I know what the body is, and I can only know what the body is when its own particular energy is fully deployed, which requires that it not be confused with the energy of any other center. So too with sexuality, feeling, thought, or whatever else we put on the list of the essential faculties or levels of the embodied soul. The disentanglement of the centers thus serves, and empowers, self-remembering and/ or self-enquiry. (There is, however, a potential down side to the mutual disengagement of the centers. The ideal is a harmonious interaction between them under the rule of the Self—but if this harmony, based on a true ontological hierarchy, fails to be established, the result can be a schizophrenic disassociation of the various functions of the psycho-physical system, leading to action without forethought, sexuality without emotional engagement, thought which ignores all questions of value, etc.)

To work directly with the energies of the centers is complex, time-consuming, and can easily lead one into the by-ways of fascination with psychic and 'occult' powers, of which Gurdjieff himself was an adept. Though this approach may be right for some, authorities from many traditions maintain that the Invocation of God's Name is the most effective practice for the times we live in, perhaps because the subtle-energy plane is in a state of chaos all over the planet, thus making the direct 'technical' manipulation of the psycho-physical energy-system increasingly dangerous and susceptible to delusions and glamors of all kinds. Ramana Maharshi himself said that 'whoever does *japam* gets realization.' Furthermore, to fully realize that 'I' (the Self) 'am not the I-sense' virtually accomplishes in one stroke what the disentanglement and deployment of the energies of the psycho-physical centers does piecemeal—though this may merely be an 'idealistic' way of putting it, given that the 'knots' created through a lifetime's identification of the I-sense with the various centers certainly do not dissolve over night, and that they may never appear in a form that can be concretely engaged with unless they are separately distinguished and activated.

The real question is, does a particular individual actually have the capacity to succeed in this practice? If you can really keep your attention on the rising and dissolving of the I-sense from the standpoint of the Self, then nothing further is needed. But if you can't, or if you practice this witnessing in an imbalanced or incomplete way such that certain elements of the soul are repressed rather than purified and transcended, then your realization will be either be ephemeral or dangerously delusive. The Invocation of the Name—a concrete technique which is simplicity itself—avoids, in my opinion, both the complexity of the Gurdjieff techniques and Ramana Maharshi's command to leap at once over all obstacles through radical self-enquiry. (Most Sufis maintain that the Invocation cannot be practiced effectively unless authorized by a *shaykh*—however, in the case of the Christian Jesus Prayer, that '*shaykh*' is Jesus Himself.)

According to a *hadith qudsi*, 'Pray to God as if you saw Him, because even if you don't see Him, He sees you.' I believe that this can be taken to refer to three distinct levels of the Invocation.

To begin with, 'I' am speaking God's Name, struggling to remain conscious of Him, or asking Him to reveal His Presence—to 'me'. The 'God' I am dealing with on this level is what Ibn al-'Arabi called 'the God created in belief.' He is inseparable from my ego; in a way He is a projection of it; contemplating Him on this level is the same thing as contemplating myself: this is the meaning of '*as if* you saw Him.' Nonetheless, as Ibn al-'Arabi points out, God still accepts the prayers we make to our image of Him, even if that image is an illusion—because, in another sense, it really *is* Him. God, as the Essence of all forms, is indeed worshipped—though in a veiled way—through the worship paid to idols, even if the idol is our own ego. According to Shankaracharya, the ego, even though it is the veil over the Self, is also a sign of the Self. It is true that those who worship God in the form of their own egos experience Him as Wrath; but if they were to allow that Wrath to take them beyond the level of personal and religious idolatry, it would be revealed as a face of Mercy.

At a later stage of the Invocation, the one represented by 'even if you don't see Him', we realize that our image of God is really only the projected shadow of our ego, and begin to encounter God as He is. At this point the words of Abu Bakr apply: 'To know that God cannot be known is to know God.' The limited, egotistical 'knower' is bewildered, neutralized; the unknowability of God consumes all our attempts to know God in the fire of Knowledge itself; this is *fana*, annihilation in God. Here we are no longer invoking God's Name; God is invoking His own Name within us. The Name no longer refers to a separate Object as perceived by a separate subject; God and His Name are One. (The intoxication of this level may in fact be what Gurdjieff called 'the opening of the higher emotional center.')

At the final stage, the stage of 'He sees you', the Invocation is the eternal arising of Allah out of His own Essence, and His eternal return to that Essence, without motion, without alteration. 'He' is *Hu*: the Self, the Divine Essence. And given that God sees nothing and no-one but Himself, 'you' is *Allah*, Lord of the Worlds, Owner of the Day of Judgment. You—as witnessed by God, not by yourself—are God's Name. The Invocation of the Name on this level is the eternal arising and annihilation of the I-sense before the face of the Divine Witness, not as the little individual self, but as God Himself. In the words of Ramana Maharshi, 'Ishwara, the personal God, is as real as your own body and mind. He does indeed create and rule the entire universe'— in the language of the Qur'an, He is 'Lord of the Worlds'—'and yet, from the standpoint of the Self, he is nothing but the final thought'—the dissolution of that final thought being one meaning of 'the Day of Judgment'. (The sobriety of this station may correspond to Gurdjieff's 'higher intellectual center'.)

'He sees you' is, however, not only the final stage of realization, the stage of 'subsistence in God' *(baqa)*, which implies one's subsistence as a Name of God; it is also virtual, and effective, from the beginning. Starting from the premise of Gurdjieff's 'self-remembering', we can say that to remember one's whole subjective self is to realize—for a moment—that the Rememberer can't be a part of that subjectivity; as C.S. Lewis pointed out, 'the knowledge of something is not one of that thing's parts.' It is to draw the energy for 'self-remembering' not from the psycho-physical centers but from the Source of consciousness itself. Once we reach certainty as to the reality of this Source, it becomes the sole Object of remembering. And once this Source is fully realized, it becomes the sole Subject of remembering as well; the limited subjective selfhood we were trying so hard to stay conscious of is first forgotten *(fana)*, and then transformed into the object of God's remembering *(baqa)*, an object which is fundamentally none other than God Himself. Conversely, starting from the premise of the Sufi practice of *zekr*, to catch a glimpse of the reality of God by forgetting yourself—or to remember God so completely that the individual self is forgotten for a moment—is to realize that you are seen more fully and more penetratingly by Him than you could ever see yourself. *Remember Me, and I will remember you.* And such Remembering in reality begins not with you, but with Him: *It is We who have sent down the Remembrance.* To remember God until God becomes the only Rememberer is to stand as the mirror of Allah, and realize the Supreme Identity.

A Tantra

Prajña

The first law of Tantra is: *Attention invokes energy; energy empowers attention.*

The archetype of attention is the *atman*; the Witness; the Absolute. The archetype of energy is Total Possibility; the Infinite. The Absolute is *Shiva*; the Infinite is *Shakti.*

Attention invokes energy, because the act of attention creates a void in the field of egotism—in the field of obsessive self-definition as well as in the secondary field of obsessive world-definition that emanates from it. *Energy empowers attention* because 'nature abhors a vacuum'. A void of egotism, or even any *relative* void, posits a space into which the world can flow as energy—as *Shakti*. Attention invokes energy.

Energy empowers attention—either that, or it creates worlds. (It is because this 'either' first arises in the field of energy, not at the point of attention, that Consciousness, as Meister Eckhart asserts, takes precedence over Being—even though, in the One Reality, Consciousness is perfectly at one with Being, with absolutely no distinction between them.) In God, the circuit 'attention invokes energy, energy empowers attention' is closed: God sees only Himself. In any form of being or field of consciousness which is less than God—in the consciousness/world of any sentient being, from the lowest to the highest—the energy invoked by attention can either empower attention—this being *vidya-maya*—or attenuate and scatter attention by creating worlds— this being *avidya-maya*. The attenuation and scattering of attention *is* the creation of worlds.

In God, perfect attention invokes total energy, and total energy is not other than perfect attention; in the One Reality, the Infinite is not other than the Absolute. But in sentient beings, attention is imperfect; therefore the energy it invokes partially flows inward to empower attention, and partially flows outward to create worlds. Given that God is the only Reality, sentient beings are embraced by that Reality; thus, within the unitary field of perfect attention united with total energy, the potentiality of imperfect attention/fragmented energy already exists. This *is maya-in-divinis.* If evil is the possibility of impossibility, existence is the possibility of limitation, the possibility of a possibility which is less than Infinite. The perfect Attention of God is perfectly able to receive the totality of His Energy, perfectly able to conceive and perceive and actualize the Totality of Possibility within Him. Yet necessarily embraced by that total attention is an attention to the discrete possibilities within the field of Total Possibility, not only attention to the infinite field of possibility itself. Given this, possibilities must emerge into existence, must be actualized conditionally and imperfectly, must be born.

Every discrete possibility within the field of total and infinite possibility is a sentient being, a consciousness/world. The attention of this sentient being (consciousness) invokes a field of energy (world). This field of energy goes partly to empower attention, and partly to make more worlds.

These worlds are veils over total and perfect consciousness; they are produced by the wavering and scattering of attention. They cannot be dispelled by eliminating them from the field of consciousness. They can only be transformed from veils into theophanies by being witnessed within the context of that field. Attention is originally perfect; it is the Witness, the *atman.* Since it appears nowhere as a possibility within the field of Infinite Energy, it cannot be diminished; it cannot be veiled. Yet the outflow of energy that creates all worlds in effect does veil It by positing sentient beings; limited witnesses; witnesses less than the Absolute Witness.

The worlds these sentient beings witness are veils. At the first eternal moment of witnessing, they know them to be veils. At each successive present instance of this eternal moment of witnessing, they either witness them or exploit them. Witnessed, they vanish. Exploited, they create worlds. A sentient being witnessing a created world has forgotten that he is witnessing a veil, forgotten that no such world has ever been created. Therefore that world has indeed been created.

The tantric path is not the path of the exclusion of illusory worlds as hindrances; it is the path of the inclusion of illusory worlds as theophanies. 'How delightful confusion is when recognized as Wisdom!' said Milarepa. Whenever a world is posited, thus positing a limited sentient being capable of perceiving such a world, a sacrifice is also posited: the sacrifice of that limited world of perception, not by excluding it as illusion and hindrance, but by including it as an instance of the Self-manifesting radiance of the Absolute: 'The conduct of the passions and attachments is the same as the conduct of a bodhisattva, that being the best conduct.' (*Guyasama Tantra*)

To attempt to exclude passion-created illusory worlds from the field of attention is to generate passion-created illusory worlds, because exclusion itself is a passion. To include passion-created illusory worlds within the field of attention is to transmute passion into impassivity and delusion into Enlightenment. It is to perfect that field. It is to transmute fragmented, partial energies into total unitary Energy. It is to synthesize many possibilities and their corresponding fields into a single field of infinite possibility. It is to dissolve possibility itself into the Infinite. It is to dedicate total energy to the perfect empowerment of attention. It is to allow perfect attention to totally embrace infinite energy. It is to reunite the Absolute and the Infinite, which have never been separated. It is to accomplish nothing; it is to recognize what is.

To capitulate to the total field of energy as conditioned by passion and ignorance, recognizing it *as* total and therefore as *unconditioned* in spite of passion and ignorance, is to worship Kali. It is to enact Shiva prostrate, mounted by the Black Goddess with keen blade drawn, mouth open wide to devour all the worlds. The infinite field of desire-energy, recognized as unconditioned, devours its own veils, limitations and conditions, including all you have held yourself to be, down to the last ripple: 'Death, thou shalt die!'

This capitulation happens when it will. The Highest Being, Ishvara, recognizes illusion the moment it arises and so sacrifices it to Truth; he recognizes Himself at the moment He arises and so sacrifices Himself to Truth. It is by this that he is Ishvara, Lord of the Worlds. By transcending all the worlds, He creates them. He creates them by allowing the sentient beings, the *demiurges* within His field of attention, to create them by believing in them. The lowest sentient beings choose, again and again, to create worlds instead of sacrificing and dissolving themselves as complexes of consciousness-paired-with-world; they follow *Mahashakti* into the abysses of dimensional existence instead of invoking Her to come back out of the wilderness of that existence to annihilate them, bringing the whole wilderness of existence with Her—not as name and form but as Energy—to destroy them as limited conscious beings posited by imperfect attention, to empower and perfect their attention, to wipe the holder-of-attention out from the total field of existence now transformed into the field of infinite Energy, to re-establish attention as Witness, to re-establish Witness as *atman*, which has never been other than established because it is That Alone by which all things are established, which is inviolable, eternal, adamantine Thunderbolt. However long they elect to hold on to themselves as separate knowers, so long lasts their *mahayuga*. As soon as they no longer elect to hold on to themselves as separate knowers, the *mahapralaya* comes.

And from the inviolable, eternal, adamantine point of the *atman*, which nowhere appears in the field of infinite possibility and universal existence, never have all these worlds been created; the dream of their creation and deployment and corruption and death—which a merciful God, Ishvara as Lord of the Worlds, sees as fully real through the eyes of those who are dreaming it—is embraced by Shiva as His Shakti, is embraced by the *atman* as nothing in itself, nothing but the field of the infinite Self-manifesting radiance of the *atman*.

And That, is That.

Upaya

On the level of concrete spiritual practice, the central question is: How can Energy be used to empower attention instead of to create worlds? What is the actual practice, the *upaya*, of this Tantra?

The practice is: to totally submit to—and therefore fully witness—all changes in psychic state.

The attempt to exclude psychic states from attention—that is, to stop the play of thought/feeling/imagination—while concentrating on the Absolute Reality beyond them is to posit a duality between Absolute Reality and thought/feeling/imagination; such duality is nothing other than the loom upon which Maya weaves the world-illusion out of the yarn of thought/feeling/imagination.

What is required is to witness thought/feeling/imagination from the standpoint of the Absolute Reality, as the Infinite Self-manifesting Radiance of that Reality. However, the attempt to witness thought/feeling/imagination from the standpoint of the Absolute Reality can only be successful if the Absolute Reality is already realized. If it is not already realized, then all attempts to witness thought/feeling/imagination from a point beyond them will only go to feed the delusory attempt to establish one element or complex of thought/feeling/imagination—one limited-self-concept, in other words—as witness of the rest of the field of which it is a part. However, the part cannot witness the whole; to treat one element of the field as it if were the Transcendent Witness of that field is the essence of idolatry.

The illusion that we in our limited ego-selves can control Maya, that we can manipulate, quash or extricate ourselves from the field of Maya, of which the limited ego-self is a part, is the essence of *avidya-maya*, of Maya-as-delusion. While we are identified with our limited ego-selves, the field of Maya, which appears within our limited, ego-bound psyches as the field of thought/feeling/imagination, is greater than us, more powerful than us, more cunning than us. We do not encompass it; we are encompassed by it.

This Maya is our prison—but it is also the Infinite Self-manifesting Radiance of the Absolute Reality. The essence of this *upaya*, therefore, is to recognize all psychic states as the Infinite Self-manifesting Radiance of the Absolute Reality, to treat them as such. How do we treat them as such? We do so by submitting to each change in psychic state, whatever its content may be, recognizing these changes to be the direct manifestation of the Will of God. Instead of seeking 'spiritual' states and fleeing from 'unspiritual' ones, we recognize all states as present Acts of God within the human soul. It is precisely this—not the Promethean attempt to reach or identify with or realize or unite with or turn ourselves into the Absolute Reality—that constitutes the unveiling of the Spiritual Heart. In the words of the Holy Qur'an, 'God holds the Heart between his two fingers, and He turns it as He will.' The Arabic word for 'heart' is *qalb*, and derives from the root QLB or QBL, which embraces a complex of meanings including 'to turn, to turn around, to overturn, to turn back, to return.' Submission to changing psychic states as Acts of God posits the Absolute as a Reality transcending psychic states, and does this in a truthful and stable way—which certainly cannot be said for the self-willed attempt to think or feel or imagine what that Reality might be, and then try to identify with It. (NOTE WELL: To understand all changes of psychic state as acts of God is not a *belief*; it is an *insight*. And this insight is available only to those who have first fulfilled all the duties God has laid upon them; in order to meet Khidr, first you have to be Moses. If man's extremity is God's opportunity, then the end of man's will is the beginning of God's Will—and no-one can come to the end of human will except by willing. Before this insight dawns—and it will normally dawn in a gradual way—we are required to deal with psychic states *as if* they were the products of secondary causes: the state of the world, past experiences, individual psychology, present circumstances, angelic and demonic influences. But we must also make ourselves ready to receive the insight that they are ultimately the sovereign acts of the First Cause alone.)

The ability to submit to psychic states as Acts of God succeeds, embraces and perfects the ability to submit to outer events as Acts of God, which normally comes first. Submission to changing psychic states as God's present actions transforms those states from veils into theophanies. It posits Absolute Reality as the Witness of those states, according to the prophetic *hadith* 'pray to God as if you saw Him, because even if you don't see Him, He sees you.' When identification with the Absolute Witness, the *atman*, is broken, then the Absolute Witness, the *atman*, is unveiled. To submit to changing psychic states as Acts of God is to know that God sees me; to know that God sees me is to unveil the Absolute Witness without identifying with it. If I believe that I am the deliberate, or helpless, author of my own psychic states, then I am using the Energy manifest in and as those states to create worlds. If I know God as the sole Author of my psychic states, then I am employing the Energy manifest in and as those states to empower spiritual attention.

The way to transform the power to create worlds into the power of spiritual attention is to submit to changing psychic states as Acts of God. To submit to changing psychic states as Acts of God is to attribute the power of creation only to God, not to myself; it is to witness God in His act of creating worlds. If I can witness God in His act of creating worlds, then I will know myself, witnessed by God, as one of those created worlds. If I know myself, witnessed by God, as a world presently created by His direct action, then I have transferred the act of witnessing from the site of the illusory ego to the site of the Absolute Witness.

And so the work is done.

Explanatory Appendix

The section *Prajña* above employs images and concepts from both Hindu and Buddhist Tantra, though the Hindu aspect predominates; it is largely a commentary on the metaphysics of Frithjof Schuon. The section *Upaya* is essentially a commentary on the teaching of the Pir of the Nimatullahi Sufi Order, Dr Javad Nurbakhsh.

Prajña means 'wisdom'; *Upaya* means 'spiritual method'; in

Vajrayana Buddhism, the marriage of *Prajña* (the feminine principle) and *Upaya* (the masculine principle) is Enlightenment. In
Hindu (Shaivite) Tantra, the union of *Shiva* (the Divine Masculine; the *Shaktiman* or 'power-holder'; the Formless Absolute; the
Transcendent Witness; the eternal symbol of the *atman*) and
Shakti (the Boundless Infinite; the field of universal manifestation both deployed as the spiritual, psychic and material universes and latent within the Formless Absolute as *maya-in-divinis*;
the field of Infinite Possibility as both the Power to create universes and the Power to return universal manifestation to its
Absolute Source) is *moksha*, Liberation. *Mahashakti* is 'great
power'. *Mahayuga* is 'great cycle-of existence', one single cycle of
four yugas or world-ages. The *Mahapralaya* is the 'great dissolution', the reabsorption of even the highest spiritual universes
into the Formless Absolute. A *hadith* is a traditional saying by one
of the founders or great figures of Islam; a prophetic *hadith* is a
saying attributed to the Prophet Muhammad, peace and blessings be upon him. *Khidr* is the name given by Sufis to the mysterious master or prophet encountered by Moses in the Qur'an,
whose actions seemed meaninglessly destructive to him until
the meaning behind them was revealed.

The Hindu *tantras*—a word which refers to both 'tantric scriptures' and 'tantric practices', and which literally denotes 'a continuous *thread* or *web*'—take Energy, Power, *Shakti*, the Divine
Feminine principle as their theoretical and practical point of
departure. In English, the concepts of *power* and *possibility* are etymologically related, as for example in the word 'might'. The possible is what *might* happen; the one who possesses *might* has the
power to realize possibilities, to transform them from potency
or potential into actuality; he is *mighty*. In Latin, *potentia* (power;
potency; potential) is unactualized possibility and stored-up
power – which is precisely how the latent *Shakti* is conceived in
the practice of *kundalini-yoga*, where the *kundalini* (*Shakti* considered in terms of the human psycho-physical nervous system) is
said to be wound up at the *muladhara* or root-*chakra* at the base of
the spine, until the Grace of God awakens it and draws it along
the path of return (the path of *laya* or dissolution-of-form) back

to its Unmanifest Source in the *sahasrara* at the crown of the skull. (The *sahasrara* is not the 'seat' of Liberation, but merely the end-point of the unfolding of the psycho-physical human form on the level of subtle energy. When this unfolding has reached its final end, then the spiritual Heart, the 'seat' or 'aperture' of the *atman*, is unveiled—not merely as the heart-chakra or *ana-hata-chakra*, the seat of affection, compassion and regard-for-the-other, situated above the level of personal power [the *manipura-chakra* at the navel] and below the level of mental or verbal knowledge [the *vishuddha-chakra* at the throat], but as the *Hridayam*, the Cave of the Heart, that void of self-definition which unveils the Eye of the Heart, that reality which the Platonists and Eastern Orthodox Christians call the *Nous*, and the Sufis, *al-Ruh*: the direct extension of the atman, the Absolute Witness.)

When *Shakti* is fully deployed, all possibility has been realized; in Aristotelian terms, Potency has been transformed into Act. God, according to scholastic philosophy, is Pure Act, since in God all possibilities are actualized, both before and after their deployment as the manifest universe. Thus the deployment of *Shakti* as the universe and the return of *Shakti* to union with its Unmanifest Source are mysteriously one and the same. In Brahman, the Absolute Reality, *Maya* is not differentiated into *Vidya-Maya* and *Avidya-Maya*; the first motion of God's manifestation is also the first movement of His veiling; He is manifested by His veil and veiled by His manifestation. And within the embrace of Absolute Reality, *Brahman* and *Maya*, *Shiva* and *Shakti*, Witness and Universal Manifestation, The Absolute and the Infinite, are *Advaita*: 'not two'.

PART TWO:

From the Lore-Stream

Hamlet's Soliloquy

A Metaphysical Exegesis

Having finally endured enough second-rate Hamlets out-herod-
ing Herod, and watched and heard even the greatest Shakes-
pearians defeated by the Soliloquy—Olivier holding irony
beneath him, though sublimity was not above him; Burton
struggling with it like an anaconda, twisting, lashing out against
the teeth of incomprehension—I've concluded that only a metaphysical exegesis can give those words the space they need to
fly, without undue torrent, tempest and whirlwind, to their
eternally-destined targets:

> *To be or not to be,—that is the question:—*
> *Whether 'tis nobler in the mind to suffer*
> *The slings and arrows of outrageous fortune,*
> *Or to take arms against a sea of troubles,*
> *And by opposing end them?*

To battle fate or submit to it, to live on or to sacrifice, or take,
one's life—of course these are embraced by the question Hamlet poses. But the question itself is bigger—so big, in fact, that
the questioner, if he be Prince Hamlet of Denmark, is too small
to ask it. Hamlet's downfall will not arise out of a wrong answer
to the question he poses, but out of the arrogation to *himself* of
the right to determine his own existence or annihilation, and
the mad belief that he has or ever could have that power. It is
God Alone who says what is to be and what not to be, not the
puny ego, not the psychic man forgetful of the Spirit, whom
Shakespeare beautifully nails elsewhere in the play by having his
hero say of himself, 'I could be bounded in a nutshell and count

myself king of infinite space—were it not for bad dreams.' The ego that labors under the delusion that it is self-created—the ego that *is*, in fact, *nothing but* this delusion—also believes that it can end itself; if this were only true, hell would hold no terrors.

The arrogant, megalomaniac ego says: To take arms is to be; to suffer is to be annihilated. The contemplative ego says: To suffer is to be, since fortune is the mask of God; to take arms is to be annihilated—because who can fight the Almighty? Who can fight the waves of the sea? Yet the contemplative ego is still an ego; capitulation to fate is still not resignation to the Will of God. What if He commands us to fight? What then? What if the only way we can be annihilated, as ego, is to cross swords with the Unconquerable? As W.B Yeats says in 'The Four Ages of Man':

> He with body waged a fight—
> But body won; it walks upright.
>
> Then he struggled with his heart;
> Innocence and peace depart.
>
> Then he struggled with his mind;
> His proud heart he left behind.
>
> Now his wars on God begin:
> At stroke of midnight, God shall win.

Only the ego is capable of making a distinction between action and resignation, self-being and self-annihilation; they are the loom upon which it is woven. The Spirit of God has never heard of them.

> To die,—to sleep,—
> No more; and by a sleep to say we end
> The heart-ache, and the thousand natural shocks
> The flesh is heir to, 'tis a consummation
> Devoutly to be wisht.

Indeed it is a consummation, and indeed it is to be wished *devoutly*; annihilation in God is, precisely, the consummation of all devotion, the end of the spiritual Path. The ego, however, cannot know this; the ego's rendition of annihilation in God is mere suicide—that great act of 'self-determination'! To the ego, death is a sleep; to the Spirit, death is the great Awakening. To die in the Spirit truly is *to die to sleep no more.*

> *To die,—to sleep,—*
> *To sleep! perchance to dream: ay, there's the rub;*
> *For in that sleep of death what dreams may come,*
> *When we have shuffled off this mortal coil,*
> *Must give us pause....*

Here the Prince begins to suspect the truth: that to be or not to be cannot be decided by the ego. And yet, what any normally religious man can do, this wise fool, this over-sophisticated, self-determined Renaissance man, this lifelong university student cannot; he cannot say: Almighty God, my life and my death are in Your hands. Too smart to believe that death ends everything, too stupid to realize—or too proud to admit—that the Almighty holds him in the hollow of His hand, what else can he do but whine?

> *For who would bear the whips and scorns of time,*
> *The oppressor's wrong, the proud man's contumely,*
> *The pangs of despised love, the law's delay,*
> *The insolence of office, and the spurns*
> *That patient merit of the unworthy takes,*
> *When he himself might his quietus make*
> *With a bare bodkin?*

Those devotees of a false self-abasement who patiently endure what is unworthy of them are spurned indeed, just as Hamlet is spurned, and beguiled, by his own unworthy philosophy.

> *.... who would fardels bear*
> *To grunt and sweat under a weary life.*
> *But that the dread of something after death,—*
> *The undiscovered country, from whose bourne*
> *No traveler returns,—puzzles the will,*
> *And makes us rather bear those ills we have*
> *Than fly to others that we know not of?*

Certainly the will is puzzled, because the will is intrinsically a servant, and in Hamlet's case—Hamlet standing precisely for modern western man—it is a servant without a Master. The

will's Master is the Intellect, the Nous, the Eye of the Heart—
which, to Hamlet, is truly an 'undiscovered country', as it must
be to anyone who tries to think with his will, who neither
directly discerns nor faithfully holds to any true Principle outside the will and its mental reflections. The human will, blind to
the Intellect or else rejecting it (and what else can the blindness
of the will be but a *willful* blindness?), can do nothing but bear
itself as a fardel, grunt and sweat under its own weariness, and
dread the void in which the Spiritual Intellect lies hidden, that
land which can only be reached when the ego dies? Truer words
were never spoken than that the spiritual traveler never returns
from that undiscovered country—and yet one Traveler did.
Hamlet, we can now see, though Prince of a supposedly Christian nation, has lost his hold on Christ, and this is a loss that all
his modern Renaissance sophistication can never redress. Christ
for the Christian is the Way, the Truth and the Life; who has
Christ has no right to call life a meaningless burden, and death
an undecipherable mystery. Death has been deciphered, and life
too; Christ is their exegesis. Yet all that remains of the Kingdom
of Heaven in Hamlet's soul is his vague but all-too-valid apprehension that insofar as the other world is undiscovered, it har-
bors nameless ills—though such ills are truly nameless only to
those who have lost the Name.

> *Thus doth conscience make cowards of us all;*
> *And the native hue of resolution*
> *Is sicklied o'er with the pale cast of thought;*
> *And enterprises of great pith and moment,*
> *With this regard, their currents turn awry,*
> *And lose the name of action.*

Thus doth 'conscience' make cowards of us all. What a telling characterization of the psychic man, the mental-willful man, to whom
the Spirit, the *pneuma*, can only appear, insofar as it does appear,
as a kind of foreknown doom. Hamlet is like a titan who foresees
that in the war against the gods the titans will lose, and so gives
up the struggle—not in resignation to the Will of God, but in
spineless capitulation to the despair of the ego, the source and

destiny of the ego's puny pride. True *conscience* is, on the moral level, firm guidance to the will, and on the Intellectual level— the level of *synteresis*—the direct apprehension of spiritual Truth by means of a 'knowing together' which marries knower and known in devout consummation—or rather, which reveals the Knower and the Known to have been, in the mystery of God's self-knowledge, of one essence from the beginning: 'Two distincts, division none: number there in love was slain' [from 'The Phoenix and the Turtle'].

Hamlet's 'conscience', however, is nothing but the mental reflection of the ego in the warped and shattered mirror of the conditional world, where nothing *must* be because nothing is intrinsically true, where anything *might* be because anything at all might happen, given that the future is an undiscovered country, harboring the ego's inevitable shadow, the grim regime of fate. Without access, through Intellection, to eternal Truth, thought is precisely *sick*, and integral action impossible. The native hue of resolution is the tincture of the Intellect impressed upon the devout and submissive will, who as a servant bears Its colors; this is *true* conscience, which makes a coward only of self-will, while conferring upon the loyal will, the will which has sworn fealty to It, unerring guidance, legitimate validation, and its true place in the cosmic order. This is the great good fortune of the will that knows its King. For Hamlet, however, the king is a usurper, just as will has usurped Intellect within his own soul; the true King, the Divine Intellect, is dead, and all Hamlet can do is follow him. Martin Lings, in *The Secret of Shakespeare*, sees a higher fealty in the prince's death, which is the death of the *ego*. When self-will dies, the prince is reunited, beyond the grave, with the true King; this is why Laertes can say, without impiety: 'Good night, sweet prince; and choirs of angels sing thee to thy rest.'

Watching
Olivier's *Lear*

Lear as King is the *Nous*, the Eye of the Heart, the One Self at the Center of all. 'The soul is an aristocrat', said Meister Eckhart. Lear uncrowned is the ego—'Lear's shadow' as the fool now calls him. Lear-as-ego believes that the love and fealty granted Lear were granted him as ego, not as King. He wants to be loved for his 'self' alone, not for his golden crown. But his golden crown *was* that Self. 'Not for their own sakes are father, child, vassal, beloved loved, But for the sake of the Self,' the Hindu scriptures declare.

The ego believes it can reign without ruling, that to be taken for a King by others—the image of a King, a player King—is to be 'every inch a King': not that the Real plays at existence, but that the play itself is real. So *Laila* becomes *Maya*. The ego veils, denies, rebels against the Self, and then expects to retain all the Self's prerogatives. But this can never be. When the crown passes from the Self to the ego, who immediately loses it, or throws it away, the faculties of the soul rebel. *Gon*-eril (lust) and *Reg*-an (the power motive) now divide the kingdom; *Cord*-elia (the Heart) is banished, murdered. Thus the Golden Age becomes senile; Manawyddan ap Llyr may no longer drive his chariot over the waves of what Blake called 'the Sea of Space and Time' as if it were a flowering meadow; Christ no longer walks upon the Sea of Galilee in the sight of men. Man falls; the Golden Age ends. This Shakespeare, living through his own lesser fall—the end of the Middle Ages in England, the late northern birth of the Renaissance, the discovery of nature, of history, of the Empire of the Ocean Sea, of man bounded in a nutshell, counting himself king of infinite space, and beset by bad dreams,

Well knew; well knew.

The Groves of Arcady

Some Templar
Correspondences

*For Roman Guelfi-Gibbs (Roman Guelfi-Gibbs is a descendent
of Dante's 'Guelphs', the son of my old Reiki practitioner, and a
Marin County deputy sheriff), who informed me that Templar
lore is filled with references to Arcadia, and conjectured that
such references, in other literature, are often a sign of Templar
influence.*

Arktos, the Greek word for Bear, is the origin of our word *Arctic*,
which is why the constellations circling the North Pole and
called the Bears; therefore *Arcadia* is one rendition of *Hyperborea*,
the land of eternal summer 'beyond the North Wind'—
undoubtedly based on reports of early Arctic explorers of the
land where summer lasts half a year, but symbolically referring
to the symbolism of the *Pole* (as in the Sufi legend of the one
holiest human being on earth at a given time, who may be
entirely unknown, but who is the true *Qutb* or *Pole* of the Age),
and the Pole Star, which is (in Eliot's phrase from *The Wasteland*)
'the still point of the turning world', the door leading out of the
cycles of manifestation and beyond the wheel of birth and
death, the visible point of Eternity in the created order. The
Siberian shamans, the traditional Chinese, the Zoroastrians, the
Sabaeans, and certain esoteric elements in Islam (see Henry
Corbin, *Temple and Contemplation*, which also has lore on the
Templars) are oriented to the North, not to the East or (like the
Greeks and Irish, at least on certain levels) the West....they are
Hyperboreans, who aspire to the *Arcadian* Paradise at the summit
of the World Axis, the path between manifest existence (the

South) and Invisible Source (the North). Dante, in the *Commedia*, is a Hyperborean in this sense; the Great and Little Bears appear above his *Arcadian* Earthly Paradise at the summit of Mount Purgatory—which is supposed to be in the southern hemisphere! (Dante is believed to have been a member of the *Fedeli d'Amore* or *Fede Santa*, which some describe as a 'third order' of the Templars.)

Guénon identifies the Templars as the guardians of the Primordial Tradition in the Christian Middle Ages. The Primordial Tradition is the original spiritual revelation of God to Man, and *as* Man, insofar as we are made 'in the image and likeness of God.' This is why the Jews and Muslims name Adam as the first prophet. Tradition knows the cycle of manifestation as a progressive departure from Source; and so the time had to come when the Primordial Tradition could no longer be known in undiluted form. This is the fall of the Tower of Babel, which represents the Promethean attempt of titanic, fallen humanity to take heaven by storm, but only after the orientation to the Pole had already been broken. The Tower, like any ziggurat, is itself a symbol of that Pole, as is every sacred mountain, sacred tree, etc. and the erect stature of humanity itself, epitomized in the 'upright' man (*tzaddik*)—all are renditions of the *axis mundi*. Muhammad (peace and blessings be upon him) was transported in his Night Journey from Mecca to Jerusalem—to the Temple Mount—from which he ascended to the higher worlds. So Jerusalem is also related to the Pole in an Islamic context, just as it was for Dante (as is, of course, the Kaaba as well).

The Templars, as Hyperboreans, knew and held to the Pole of all the religions. But since religions cannot be mixed in the world of form without polluting them and invoking demonic forces, their lore was *arcane*, to be kept hidden in an *ark*, such as the First Temple once held. *Arthur* too was a Hyperborean; his name comes from the Gaelic *arth*, bear, cognate with *arktos*. (The Pole with its two Bears was the *Caer Sidi*, the revolving castle of the Byrthonic Celts.) Among the ancient Celts, the knights (corresponding roughly to the Hindu *kshatriyas*) were *bears*, while the Druids (corresponding to the *brahmins*) were *boars*. *Bear* and *boar* are related words; Hyperborea was first the land of the *boar*—but when the *kshatriyas* revolted against the *brahmins* in parts of the Indo-European world, it became the land the *bear*. Both words are related to the Sanskrit root *var*, which means something hidden or enclosed. The Pahlevi word *var* means 'compound; enclosure'. The Var of Yima, the first man in Zoroastrian mythology, who was also the first prophet, seems to be identified in the *Vendidad* with the Pole—and, by extension, with constellations of the Bears; yet it is also situated beneath the earth. (The Var of Yima and the Cave of the Sleepers mentioned in the Qur'an are undoubtedly two versions of the same primordial idea.) The Kaaba ('cube') is, on one level, a symbol of the earth—the Platonic solid for the earth-element is the cube— and, on another, of the Pole. The synthesis of these two meanings is to be found in the Sufi doctrine of the spiritual Heart— the point where the 'Pole', the ray of Spirit, intersects the 'earth', the human psycho-physical entity. *Ark* means 'a box; something put in a box'—something which is therefore hidden or *arcane*. Going further afield, and employing what most etymologists would think of as a spurious or folk derivation, the Primordial Hyperborean Arctic tradition is also *archaic*, stemming from the Original First Principle, the *arche*.[2]

In *Symbols of Scared Science*, Guénon has much to say about the symbolism of the *cornerstone*, traditionally identified with Jesus, which in the Gospels is 'the headstone of the corner' which the builders rejected, as the Jews, the builders of the Temple, rejected Christ. (He is careful, however, to identify the 'headstone of the corner' not with the cornerstone of a foundation but with the capstone of an arch.) Guénon connects it with the philosophers' stone, with the Kaaba and its Black Stone, with the Jerusalem Temple and thus, by extension, with the Templars. (In the most Templar-influenced of the Arthurian romances, the *Parzival* of Wolfram von Eschenbach, the Holy Grail is not a cup but a stone.) 'Headstone of the corner' in Latin is *caput anguli*, the 'head of the angle'; in Arabic this phrase is rendered as *rukn al-arkan* ('angle of angles'), which can be translated as 'summit'. The five *arkan* (angles) in certain strands of Islamic esoterism are the major principles or *arcana* of the cosmos, and are related to the five elements; the 'summit' angle corresponds to ether, the quintessence.

If the four elements are angles, crowned with a fifth 'summit' angle, we have the image of a pyramid, and thus are square within Masonic lore. Guénon saw the Masons as a valid, though largely degenerate, initiatic organization, perhaps originally a Sufi order. He believed that Masonic symbolism was metaphysically accurate, but no-one can deny that the Masons have often lent themselves to beliefs and activities which were both spiritually and politically subversive. The Masonic myth of human *edification*—as in our word 'edifice'—of the building of the Temple of Man, may have been true and effective in a context which recognizes God's grace as the primary agent, but outside such a

2. Few etymologists seem to realize that 'folk etymology' is not always simply inaccurate—that, in other words, it may not be folk etymology at all. For many thousands of years, according to one hypothesis, esoteric groups have made deliberate 'puns' upon words from different languages in order to transmit arcane lore, as if in an attempt to partly compensate for the 'confusion of tongues' following upon the fall of the Tower of Babel—and a deliberately-constructed 'bridge of meaning' between two otherwise unrelated words becomes indistinguishable from 'valid' etymology after thousands of years. In a shorter time-frame, consider the word *Sufi*, which is usually derived from the Arabic words *swf*, 'wool', or *safa*, 'purity'. Yet it certainly echoes the Greek word *sophia*, 'wisdom', and the Aramaic word *suf*, which means 'reed'. When Jesus said, referring to John the Baptist, 'who did you go out into the desert to see? A reed shaken by the wind?', he was pointing in a hidden way to John as a 'hollow reed' who only spoke when and as the 'wind of the Spirit' commanded; compare the first lines of Rumi's *Mathnavi*, 'Listen to the reed, to the tale it tells of separation. . . .' It is my belief that when the Sufis, due to the downward course of the Islamic tradition, had to adopt a name at one point to distinguish themselves from the more exoteric believers, they deliberately chose a word that would resonate meaningfully in several languages. On a deeper level, however, the correspondence in form and meaning between words with no demonstrable etymological kinship is explained by the Hindu principle of *nirukta*, which Guénon in *Man and his Becoming According to the Vedanta* defines as 'the science of interpretation chiefly based on the symbolical value of the elements out of which words are built up.' The Sanskrit *Purusha*, for example, means 'Person', but it is symbolically—not etymologically—related to the word *puri-shaya*, 'city'.

context it becomes Promethean, making the Tower of Babel a more appropriate symbol of later Masonry than the First Temple at Jerusalem.

Here are the *groves of Arcady;* not dead, but hidden in the Var of Yima—the Cave of the Heart.

Chessmen
and Hexagrams

I know next to nothing about chess, but I do know something about the *I Ching*. Based on that knowledge, I believe I can present some persuasive evidence that chess actually originated in China, as certain scholars now believe.

The 64 squares of the chessboard have been identified by some with the 64 hexagrams of the *I Ching*. But it seems equally significant that just as each hexagram is composed of six lines, so in chess there are six classes of chessmen.

The third and fifth lines of every hexagram represent centers of authority. The fifth line is that of the king, while the third represents something on the order of a feudal lord. The second and fourth lines are the active servants of the third and fifth lines respectively: the fourth is the king's minister, while the second is often compared to a popular leader or a general in the field. The first line is that of the common people, the sixth and last that of the sage, who may either withdraw into retirement or remain in the world as adviser to the king.

In view of these attributions, I would say that the first line of a hexagram (the common people) corresponds to the pawns, the second (the general) to the knights, the third (the feudal lord) to the castles, the fourth (the minister) to the bishops—who so often functioned as king's ministers in European history—the fifth to the king, and the sixth (the sage) to the queen—a piece which, in Persia, is named not 'the queen' but 'the adviser'. Since these correspondences are far more specific than could easily be explained by chance, I see them as further evidence that chess did in fact originate in China, and is based on the structure of the *I Ching*, certainly one of the most ancient books we possess, and one which existed as a system of divination based on an orally-transmitted cosmological system long before it was committed to writing. (It is said that the 64 hexagrams were suggested by the patterns made by an empty tortoise-shell as it cracks in the heat of the sun. Is this in fact also the origin of the chessboard?)

Speculations on
Esoteric Ecumenism

In René Guénon's
Insights into Christian
Esoterism

On pages 43–44 of *Insights into Christian Esoterism*, Guénon, follow-ing Luigi Valli, and within the general context of the Templars, mentions the peregrinations of initiates as forming a hidden aspect of pilgrimage. Were such esoteric pilgrims to be found among the Templars? Was the founding of the Order of the Temple, as an armed response to the threat posed to the pil-grimage routes by the Muslim conquests, partly inspired by such pilgrims? Guénon identifies the Templars as among the 'Guardians of the Holy Land' of the primordial and unitary Tra-dition, in relation to the assertion of the figure of Melchizedek in the *Decamerone* of Boccaccio—named as among the Fedeli d'Amore—that no one knows whether Judaism, Christianity or Islam is the true religion.

Could the Templars have occupied the Temple Mount in Jerusalem partly to prevent the other crusaders from destroying the Dome of the Rock, thus combining the exoteric role of Christian holy warriors against Islam with the more esoteric function of preserving Jerusalem as a point-of-origin for the unitary Tradition by preventing it from being *monopolized* by Islam? The report of Usama ibn Munqidh, Emir of Shaizar, that the Templars used to put an oratory attached to the Dome of the Rock at his disposal so that he could pray, and defended his right to do so against Frankish 'newcomers' (cf. Edward Burman,

The Templars, Knights of God, 1986, Destiny Books, p. 76) can be explained by the fact that it would have been equally in line with this (conjectural) agenda for the Templars to prevent the sanctuary from being monopolized by the *Christians*. Nothing would have more effectively cloaked this purpose than the Templar presence as part of the Christian army of occupation! It might also explain their recklessly brave but often strategically unsound war-making: better suffer high casualties than ever be suspected of an 'esoteric' collaboration with the enemy. (Making obeisance to Saladin was one of the charges brought against them at their trial, centuries later).

On p. 51, Guénon writes:

> Another important point concerns the relation between the Fedeli d'Amore and the alchemists. A particularly significant symbol in this regard is found in Francesco da Barberino's *Documenti d'Amore*. The figure in question consists of twelve personages arranged symmetrically and forming six couples which represent as many initiatic degrees, surrounding a single figure at the center; this last, who holds in his hands the symbolic rose, has two heads, one male and one female, and is manifestly identical with the hermetic *Rebis*.

Compare this with the following passage from 'The Wish of Manchán of Liath' by an anonymous Irish monk of the 10th century, from *A Celtic Miscellany*, translated by Kenneth Hurlstone Jackson, Penguin Books, 1971, p. 280:

> I wish, O son of the Living God, ancient eternal King, for a secret hut in the wilderness that it might be my dwelling....
> [and for] A few sage disciples, I will tell their number, humble and obedient, to pray to the King.
> Four threes, three fours, fit for every need, two sixes in the church, both south and north.
> Six couples in addition to myself, praying through the long ages to the King who moves the sun.

The central androgynous figure in Barberini's book on the Fedeli d'Amore, holding the rose, is obviously Love; he is the Hermaphrodite, the union of Aphrodite (love) and Hermes (knowledge); compare the feminine figure of Condwiramurs in the *Parzifal* of Wolfram von Eschenbach, the meaning of whose name (according to Henry and Renée Kahane, Eschenbach's greatest exegetes) means 'knowledge of love', as well as the figure of Dante's Beatrice, in whom Wisdom and Love are united. Thus 'the King who moves the Sun' named by the Irish monk, perhaps four hundred years before Dante, is almost certainly related to, if not identical with, that image of the Deity with which Dante's *Paradiso* concludes, 'the Love that moves the Sun and the other stars.' True affinities may or may not be historical: they are always intrinsic, and providential.

An Exegesis of the Prologue to William Blake's *The Marriage of Heaven and Hell*

Inspired by a Visit to Glastonbury, with an Appendix on the Legend of the Grail

I am an American poet whose bloodlines flow mostly back to Britain. On a recent visit to Somerset, guided there by my British host and hostess who dearly love the place, my British ancestral mythopoetic unconscious was stirred and opened:

Rintrah roars and shakes his fires in the burdened air

Rintrah, according to S. Foster Damon, is the sign of God's wrath in Blake's mythology.

Hungry clouds swag on the deep.

The clouds are ships of the Royal Navy under full sail, 'swaggering' on proud and warlike missions for the Empire; these first two lines are allusions to the atmosphere of the times in which Blake was writing, times of the American Revolution, the British counter-revolution (which led in England to popular protests much like those in America against the Vietnam War), the French Revolution, and, in reaction to it, the suppression of domestic liberties; Blake himself was once tried for sedition.

Once meek and in a perilous path,
The just man kept his course along
The vale of death.
Roses are planted where thorns grow,
And on the barren heath
Sing the honey bees.

The monks at Glastonbury, where the Thorn grew, and who (as I learned from a friend of our hosts, who served us her rose nectar made according to a monkish recipe) cultivated roses (sacred to the Virgin Mary, the patroness of Glastonbury Abbey), kept to the straight path of salvation, the path of self-annihilation, of death-before-death, in the Vale of Avalon, which, as the Celtic otherworld, is also the vale of death. And monks, like·bees, live in *cells.* I myself saw beekeepers in Glastonbury.

Then the perilous path was planted,
And a river and a spring,
On every cliff and tomb
And on the bleached bones
Red clay brought forth.

The monks were martyred and planted in their graves, planted only to sprout up again in resurrection—and the bloody spring of Glastonbury (the water so *chalybeate,* so full of iron, that it tastes like blood) is their martyr's blood springing from the ground, reddening (as I saw with my own eyes) with iron rust the little cliff over which it pours, then flowing into Arthur and Gwenivere's tomb—which is directly down hill from the spring—who represent all the dead of England, and the Earth too—in aggregate, Adam the Primordial Humanity, whose name means 'red clay'. The vivifying blood of the martyred monks raises Arthur, who is Adam (and thus a type of Blake's Albion) from his tomb, clothing his and Gwenivere's bones in living flesh. (There is strife in that tomb; I fear for the day it is opened.)

Then the villain left the paths of ease
To walk in perilous paths, and drive
The just man into barren climes.

If 'the perilous path' is the path of religion, then it was King Henry VIII who took the perilous path of declaring himself pope of the English church, and driving the monks, the just men, from their monasteries.

Now the sneaking serpent walk in mild humility

The priest of the Church of England has now abdicated his spiritual function and become the propagandist for British imperialism, the hypocritical spirit who fills the sails of those 'hungry clouds'.

And the just man rages in the wilds where lions roam.

The Holy Spirit, having abandoned King and Church, has gone into the social wilderness to inspire marginalized and wrathful prophets—like Blake himself.

But the resurrection of Arthur is ambiguous, in line with Blake's identification of the Druid religion (Arthur's mentor was Druid Merlin) with vengeance and political oppression, as opposed to the reign of Christ, based on the forgiveness of sins. So if the return of Arthur is the return of Primordial, Adamic Man, it is also the revolt of pagan Druidism and the cult of the warrior-king against the 'just men' of Christianity—a revolt personified by King Henry, which has ultimately led in our time to an almost complete capitulation of the Anglican/Episcopal church to Neo-Paganism, including even (in America at least) witchcraft. (It is really the standard-bearer in this regard. Not only is England the most secular nation in history, but the Episcopal Church in the U.S., according to a recent survey, has the largest number of *atheists* of any Christian denomination.)

Well are Blake's books called Prophetic! When, in that lyric from his *Jerusalem*, 'And did those feet in ancient time/Walk upon England's mountains green?', alluding to the legend of Joseph of Arimathea, a tin merchant and uncle of the Virgin Mary according to legend, who is said to have taken the boy Jesus to England on one of his trading journeys, landing near the mines of Somerset, the Vale then being an inlet if the sea—Blake said: 'I will not cease from mental fight/Nor shall my sword sleep in my hand/Till we have built Jerusalem/In England's green and pleasant land', he was, in a way, declaring himself a one-man non-Roman Catholic esoteric English counter-reformation.

And the Grail

The Chalice Well at Glastonbury, with its water that tastes like blood, is, precisely, the Holy Grail. The site was reputedly the earliest Christian site in Britain, founded by Joseph of Arimathea, who, according to legend, brought the Grail containing Christ's blood to Glastonbury after the crucifixion; the monks would readily have identified the chalybeate water with the blood of the Savior, who is sometimes called a Fountain of Living Water.

The Chalice Well incarnates the lesser, feminine, psychic mysteries, the return to 'Adamic' state, to the human essence as God created it, the realization of the Earthly Paradise. The Tor on the hill directly above, sacred to St. Michael, on St. Michael's Ley, is the Vertical Path, the *axis mundi*, a ray of the greater, masculine, Spiritual mysteries which lead to the transcendence of the human state, to Union with God. (First Lethe, then Eunoë; first the Earthly Paradise, then the *Paradiso* itself.) St. Michael in his icons is most often pictured carrying a lance—and so the mystery of the pairing of the Grail with the ever-bleeding Lance in the Grail romances is no mystery: the blood which mingles with the pure water of the Chalice Well ultimately comes from Above; the spear of Longinus piercing the side of the crucified Christ, the lance of St. Michael, and the cross itself (like the Tor, and like Blake's engravings of *Jacob's Ladder* and *The Last Judgment*) are renditions of the *axis mundi*, the path which unites the created universe with its unseen Source.

And one more thing: On the border of the Orthodox Christian icon of Our Lady of Glastonbury (who is overshadowed by a smaller figure of St. Michael, bearing a lance, and who bears in her right hand the Thorn and in her left the Christ Child, himself carrying a globe of the heavens) are represented many ancient, local Christian saints, one of whom is *St. Kea*. Upon my return to California I consulted Eastern Orthodox lay nun Katherine McCaffrey, a trove of spiritual and historical lore, about who these saints were, and encountered the story in one of her books that St. Kea was King Arthur's chaplain, who packed Gwenivere off to a nunnery after her adultery with Launcelot and the dissolution of the Round Table. He was Arthur's *staretz* (Russian) his *geron* (Greek), both of which mean 'elder'. So he was undoubtedly the same figure as *Sir Kai* of the Round Table, Arthur's *seneschal*, which also means 'elder' or 'old man'. Kea is the Church image, and Kai the knightly or warrior image, of the same man. Sir Kai was a foul-mouthed, vain and curmudgeonly older knight, threatened by the prowess of younger and stronger men; imagine this as the picture keen young warriors would have had of a pious, admonishing, older Christian monk—St. Kea—in a time when the Christianization of Britain was far from complete.

Homer,
Poet of *Maya*

In his great epics the *Iliad* and the *Odyssey*, Homer expresses a complete doctrine of *Maya*, and of the cycle of Divine manifestation as unfolded and dissolved by *Maya*. Taken together, the two poems constitute a mythopoetic *cycle* in the strict sense of that term: the story of the creation and destruction of the universe (a word that means 'one turn')—which, in sober fact, is the only story there is.

According to the Hindus, Maya is the Great Mother of manifest existence; She is the creative illusion or magical apparition of Brahman, the Formless Absolute, in terms of the finite field of forms. Maya is not strictly unreal or non-existent; it's simply that She is not what She seems; to use the traditional Hindu simile, She is like 'a rope mistaken for a snake'. Maya appears in two forms: *avidya-maya* or 'ignorance-apparition', and *vidya-maya* or 'wisdom-apparition'. Since Maya is not what She seems, she seduces us to try and make sense of Her, a task which is both inescapable and ultimately impossible. As *avidya-maya* she lures us into a false identification of Absolute Reality with the relative world, and in so doing creates that world; as *vidya-maya*, she lures us (as in Plato's *Symposium*) toward identification with ever higher and more comprehensive images of Absolute Reality, each of which is progressively discarded in favor of a greater conception, until all images and conceptions are finally transcended and Absolute Reality realized.

The *Iliad* is the epic of *avidya-maya*, the *Odyssey* the epic of *vidya-maya*. Helen, the cause for an expansive, imperialistic war, the 'face that launched a thousand ships', is *avidya-maya*, the power that lures men into worldly identification, conflict and dissipation. Penelope, the wife of Odysseus 'the cunning', the image of retreat, of withdrawal from the world, of home, is *vidya-maya*. She is the power of recollection, of return to the 'center in the midst of conditions'; she is of Holy Wisdom. In the epics of Homer, Troy is the City of This World, and Ithaca is the City of God.

The end of the expansive and dissipative attempt to 'conquer the world', to control material conditions, to possess and dominate the 'ten-thousand things' is, precisely, *apocalypse*—the end of the world—the burning of Troy. And Odysseus knew this. Like any wise and cunning man, he knew that the end of the war to conquer the world would be destruction and nothing else, that Troy would burn, that the victorious Agamemnon would be murdered—and that Helen, once 'rescued', would simply become irrelevant. And so, of course, he tried to avoid the conflict; he flinched at the call to be born into this world. He, like many 'draft dodgers' in the Vietnam War, feigned madness to avoid service; he tried to plough his field with his plough harnessed to an ox and an ass, expressing the inevitable divergence of intent and division of the will that this world is made of, due to the fragmentation of the original human character (as the giant Ymir, in Norse myth, was dismembered to create the universe). But when his son was placed in front of the plough, he turned it aside—proving that he was not mad enough to destroy his own spiritual center and destiny. And so he had to go to war. He was wise enough to foresee the inevitable tragedy of the fall into this world of division, conflict and destruction—but Maya, the deceiver, was wiser than he. Only She knew the darkest secret of existence: that the descent into this world is really a *felix culpa*, a 'fortunate sin'; that in the depth of the Great Mystery, the loss of God—felt and fought against and suffered through and finally redeemed—is in fact the deepest realization of God, that after 'My God, my God, why hast Thou forsaken me?' comes 'Into Thy hands I commend my spirit; it is finished.'

Helen, in the *Iliad*, is shown sitting in Priam's palace in Troy, weaving a 'double purple cloak' upon which appear the scenes of the great war between the Trojans and the Achaeans that she herself precipitated; by this she is revealed precisely as *avidya-*

maya, who weaves the pattern of manifest existence. But Penelope, Odysseus' wife, is also a weaver. She has told her many suitors that she will wed one of them when the shroud she is weaving, the shroud of her father-in-law Laertes, is done—but every night, in order to put them off, she unweaves what she had woven during the day. Laertes, father of Odysseus, is his Principle and Origin, the aspect of God that pertains most directly to him, his archetype *in divinis*. And this world, to the wise, is not a cloak of the living, a purple garment fit for kings, depicting glorious and heroic struggles, but, precisely, the shroud of God, the veil that covers Him in death. This world unfolds in all its convincing multiplicity only when God is dead to us; only then, under the influence of *avidya-maya*, do we mistake this perishing world for Reality itself. Helen weaves the cloak of the world-illusion, but only Penelope shows it for what it is. Only *vidya-maya* can reveal to us the secret of Maya per se, that the pattern of existence woven on the Day of Brahman is unwoven again in the Night of Brahman, that the world created by Maya is not a stable reality, but a coming and going, a wheel of birth and death, an outbreathing and inbreathing of the Great Sleeper who dreams the universe.

The course of *vidya-maya* embraces the many journeys and battles and awakenings and realizations that are the spiritual Path; the *Odyssey*, the epic that depicts the struggles of Odysseus to be reunited with Penelope, with Holy Wisdom (under the guidance of the goddess Athena, the active manifestation of that Wisdom, and Hermes her emissary), is the story of that Path. The stations of Odysseus' spiritual journey, symbolized by his long return from Troy to Ithaca, are as follows:

I

The raid carried out by Odysseus and his men on the Cicones of Ismarus, in which they are defeated. This symbolizes the defeat of worldly ambitions, the proof that this world no longer holds anything of value for those who have embarked upon the spiritual Path.

II

The land of the Lotus Eaters. This station, under the metaphor of drug addiction (possibly to opium), symbolizes the overcoming of World Trance, the nearly universal addiction of the human race to the sort of earthly experience that denies any possibility of spiritual experience—either that or the habit of unconsciously translating the dawn of true spiritual experience back into a different kind of worldly experience, notably the intoxicating complacency represented by the Buddhists as *Deva-Loka*, the realm of the long-lived gods where 'ignorance is bliss'. In terms of the spiritual Path, Vajrayana practitioners are warned not to fall into 'the beautiful Hinayana peace', into the complacent illusion that Enlightenment has already been achieved.

III

The Island of the shepherd Polyphemus, the one-eyed Cyclops, who devours several of Odysseus' men. Jesus said 'if thine eye become single, thy whole body will be filled with light'; Polyphemus represents the dawning of this truth, the truth of the Absolute One, on too low a level, resulting in the 'absolutizing of the relative', the error that creates religious fanatics and then devours them. The premature identification with the Transcendent One at then expense of Its multiple manifestation results only in destruction; the rigor of Transcendence can be safely encountered only after the soul has been unified in submission to God's will. Before that time, the Eye of the One—like the 'third eye' of Shiva, the opening of which destroys the world illusion—must be blinded, veiled; otherwise it will burn up all the traveler's spiritual potential instead of putting it to effective use. But what appears to be only a disaster is in fact the secret beginning of that soul-unification. Throughout the Odyssey, Odysseus' followers are gradually killed off, until only he remains alive; this symbolizes the process of *recollection*, the mortification of the various divergent impulses of the soul, in terms of the affections, the will and the thinking mind, until all that remains is

unity of character and one-pointedness of spiritual attention
and intent. As it says in the *Tao Te Ching*, 'Knowledge is gained by
daily increment; the Way is gained by daily loss—loss upon loss,
until at last comes rest.'

IV

Aeolus, the god of the winds, gives Odysseus a bag containing all
the winds of the world, so that he will always have a fair wind in
his voyage back to Ithaca. But his followers, out of curiosity,
open the bag while Odysseus is asleep, let the winds loose, and
all their ships are blown off course, back to their starting-point.
This represents a wrong relationship to Divine Providence,
based on lack of trust in God. If we dig up the seed every day to
see if it is growing, the plant will never mature; if we arrogate to
ourselves the right to see into the mind and Spirit of God, so as
to better understand all the ramifications of our spiritual
destiny—as if we could be better guardians and administrators
of that destiny than God Himself—then we will be blown far off
course. While Wisdom sleeps, curiosity wakes up and starts
looking around; in the words of Frithjof Schuon, 'mental passion pursuing intellectual intuition is like a wind that blows out
the light of a candle.'

V

Landing on the Island of the Laestrygones, who are cannibals,
Odysseus' men encounter a young girl, daughter of the king of
the island, who invites them to her father's court, where many
are devoured. The princess symbolizes *avidya-maya*, who lures
men to eat themselves up in worldly pursuits; spiritual curiosity
ultimately results in a regression into worldliness. Nonetheless,
the great purification continues. Cannibals are 'self-eaters';
under the hidden influence of the Spirit, the divergent tendencies of the soul begin to destroy themselves by their own folly.

As William Blake said, 'If the fool would persist in his folly he would become wise.'

VI

Next Odysseus and his men land on the island ruled by the witch Circe, who—weaving at an enormous loom—is Maya incarnate. She transforms Odysseus' men into pigs, symbolizing their total defeat at the hands of the lower passions. But Odysseus, the center of the spiritual Heart, is not overcome. Hermes provides him with a 'sobriety drug', an 'anti-illusionogenic' called *moly*, which is the exact opposite of the drug taken by the Lotus Eaters. It is this that allows him to overcome Circe's spells and turn his pigs back into men. And precisely *because* Odysseus resists her charms, Circe falls in love with him. This is the great enantiodromia, where *avidya-maya* is changed into *vidya-maya*; Circe is here transformed from the *maya* that deludes Odysseus into the *shakti* that empowers and serves him. From now on she is not an enchantress, but a guide.

VII

Circe now sends Odysseus to the western edge of the world, the limit of manifestation beyond which nothing remains but the One, where he invokes the shade of the seer Tiresias. Tiresias had been transformed into a woman by the goddess Hera for a period of years, for the crime of striking a pair of copulating snakes with a staff. Later she strikes him blind as well, but gives him in return the gift of foresight. The striking of the snakes with the staff invokes the caduceus of Hermes, who also has an hermaphroditic aspect. The caduceus represents the power to unite opposites, as the *ida* and the *pingala*, masculine and feminine psychic currents (the two snakes), are united by being woven around the central *sushumna*, the *axis mundi*, in the practice of kundalini yoga. Tiresias, having been changed from a man into a woman and back again, is beyond the pairs-of-opposites which are the warp and the weft of the world illusion. He is blind (like Homer was) to the multiplicity of this world, but his gift of foresight shows him the final end of the spiritual Path. He advises Odysseus on how to travel that Path to reach the final goal.

After Tiresias, Odysseus speaks with the shade of his mother, who reveals to him the plight of his wife Penelope, beset by unwanted suitors in Ithaca, and he also encounters the shade of Agamemnon, who tells him of his own murder at the hands of his wife Clytemnestra. Here the necessity of completing the spiritual Path is revealed, along with the final destiny of all who fail to complete it: division and death, in a world woven on nothing but division and death, the world of the double cloak.

Odysseus and his men return to Circe's island where she advises them further on the journey ahead, and they set sail again.

VIII

Odysseus and his men encounter the Sirens, symbols of the delusive and destructive side of spiritual Beauty, whose beautiful songs lure sailors to death on the rocks. His men put wax in their ears so they will not hear the Sirens' songs, but Odysseus asks to be tied to his ship's mast—the *axis mundi* again—so as to be able to hear the songs, while being restrained from following them. The lesson here is that only those who have fully attained the centrality of the human form can witness the Beauty of God and the esoteric secrets of the spiritual Path without being led to destruction. The Beauty and Mystery of the Divine are gifts that come in their own time; to run after them and try to grasp them, like the lustful brave who wanted to rape the beautiful emissary of God to the Lakota, White Buffalo Cow Woman, is to be destroyed by the Majesty of God.

IX

Next the voyagers encounter Scylla, a many-headed sea-monster, and Charybdis, a whirlpool. The whirlpool is *sangsara* (often compared to a whirlpool by the Buddhists), the engulfing and obliterating power of the world of relativity and formal manifestation; the sea-monster is the division of the soul—mind, affections and will—between many worldly concerns, each of which takes a piece of us. Odysseus chooses to brave Scylla instead of falling into Charybdis, losing only six men instead of the entire ship, demonstrating that even though the struggle with worldly necessity wounds us, to ignore it is fatal, and that to pass beyond the pairs-of-opposites is not done by failing to distinguish between them, but by choosing always 'the lesser of two evils'. One does not transcend good and evil by treating them as if they were the same thing, but by always choosing the good, at whatever cost, until the Sovereign Good is won—that Good which lies is beyond the opposition of good and evil because, being all good, it has no opposite.

X

The voyagers now land on the Island of Trinacis, where—contrary to the advice of both Circe and Tiresias—Odysseus' men hunt and slaughter the Cattle of the Sun. In punishment, all are drowned in a shipwreck except Odysseus, who is washed up half dead on the island of the goddess Calypso, where he must remain as her lover for seven years. The Cattle of the Sun represent a glimpse into the Majesty of God, producing in a spiritual exaltation that the ego attempts to appropriate, which ultimately results in a titanic inflation and the resultant fall. Odysseus is back in the clutches of this world—the nymph Calypso. Nonetheless, though he seems to have been defeated, the merit he gained through his earlier spiritual victories is still working in secret.

XI

Hermes now appears, and convinces Calypso to let Odysseus go. So he builds a raft and sails to the island of Scherie. He is washed up on the beach exhausted, and there encounters the princess Nausicaa, symbol of *vidya-maya*. She introduces him to her parents, the rulers of the island, and when he tells them the story of his journey they decide to help him. With their aid he returns to Ithaca, disguised as a *swineherd*—as one who has the power to control the impulses of his lower self.

XII

Odysseus, in Ithaca, is recognized by his old housekeeper due to a scar on his thigh he received in his youth from a wild boar. He enters his palace, still in disguise. The next day Athena prompts Penelope to issue a challenge to the suitors: to string Odysseus' bow and shoot through the holes of twelve axes placed in a row (These holes are sometimes called 'helve-holes', as if they were holes in the hafts of the axes by which they could be hung up, but I see them as the semi-circular spaces between the two upward-curving corners of the Cretan double-axe.) None of the suitors can even string the bow, but Odysseus can; he shoots an arrow through the holes of the twelve axes and wins the contest. He kills all the suitors, as well as the twelve maids who slept with them. He reveals himself to Penelope, who is uncertain of his identity until he describes the bed he built for her when they were first married. They are reunited.

The axes through which Odysseus shoots are double-headed axes with crescent blades, representing the waxing and waning phases of the Moon, the waning phases symbolizing the fall into the darkness of this world, under the power of *avidya-maya*, and the waxing phases the stations of the spiritual Path, by the power of *vidya-maya*. According to the *Bhagavad-Gita*, darkened souls destined for rebirth enter the waning Moon after death, while sanctified souls destined for Liberation enter the waxing

Moon, and from there pass on through the Door of the Sun. And in Sufi symbolism, the double axe represents the cutting off of attachment both to this world and the next, leading to the realization of God in this very life.

The twelve axes, which are twelve moons, are a *zodiac*, an entire cycle of manifestation, like the twelve gates of the Heavenly Jerusalem and the twelve stations of the Odyssey itself. That the twelve maids who slept with the suitors are killed symbolizes a passing beyond the sphere of the Moon, the cycles of nature and rebirth, and a realization of full Enlightenment. The number of Penelope's suitors is 108, a sacred number in Hindu and Buddhist lore. The suitors symbolize the level where worldly multiplicity is always seeking the blessing of spiritual Unity, perpetually struggling to possess the One without first becoming the One—an impossible task. When this level is killed, only Unity remains. The pre-eternal Unity, symbolized by the marriage-bed of Odysseus and Penelope, that held sway before Principle and Manifestation were polarized, is re-established; the spiritual Path is complete.

This is the spiritual principle behind all Greek civilization; this is secret that Orpheus knew.

PART THREE:

Physics and Metaphysics

We Are the
Bees of the Invisible

Physics, Metaphysics,
and the Spiritual Path

For Scott Whitaker, 1952–1998; mathematician, Platonist, fakir of the Maryamiyya Shadhili Tariqa, Orthodox Christian at the hour of his death, unacknowledged (even by himself) rediscoverer of Pythagorean mysticism, and dear spiritual friend, who transmitted to me, in his last years, in the course of our many long conversations, across the language barrier separating his profound mathematical knowledge from my ability in the field of English prose, the essence of the last seven paragraphs of this essay, which precisely complete my own cosmological vision, received when I was eighteen years old, in the summer of 1967.

May our collaboration continue until we stand, all speculation at an end, face-to-face with the Object of our quest.

.... our task is to stamp this provisional, perishing earth into ourselves so deeply, so painfully and passionately, that its being may rise again, 'invisibly', in us. We are the bees of the Invisible.... Transitoriness is everywhere plunging into a profound Being.... The earth has no other refuge except to become invisible: in us, who, through one part of our nature, have a share in the Invisible ... only in us can this intimate and enduring transformation of the visible into an invisible no longer dependent upon visibility and tangibility be accomplished, since our own destiny is continually growing at once MORE ACTUAL AND INVISIBLE within us.   *—from a letter of Rainer Maria Rilke, 1925*

It is common nowadays for many to imagine that the universe, in line with progressive and evolutionary ideas, must somehow be advancing spiritually. If we come to the conclusion that the spiritual evolution of the macrocosm is not possible, we may even wonder what is the worth or profit in material existence. What good is it? What is it for? For fear of becoming 'Gnostics' who deny the value of terrestrial life, we end by denying the *eternal* significance of this very life.

The problem with the concept that the universe evolves to higher levels of organization, which is basic to the doctrines of Teilhard de Chardin, Rudolf Steiner, and many other New Age teachers (as well as to the attempt within Judaism to apply Lurianic Kabbalah—and within Ismailism, to apply the 'unveiling' of spiritual realities—to historical evolution) is the Second Law of Thermodynamics. This law states that, via entropy, the overall order of matter/energy in the universe is always decreasing, a decrease which is inseparable in principle from the expansion of the universe, starting at the Big Bang. Furthermore, evidence is mounting that the universe is not only expanding, but expanding at an ever-increasing rate, ultimately resulting in a 'Big Rip': the moment when the expansion reaches the speed of light, and the universe is annihilated. This is strictly in line with the Hindu doctrine of the entropic nature of the *mahayuga* or cycle-of-manifestation—which, according to René Guénon, results in a progressive acceleration of time as the cycle moves toward dissolution.

'This whole world is on fire,' said the Buddha. 'All is perishing,' says the Qur'an, 'except His Face'. Creation, in the traditional view, is a successive 'stepping down' of a higher orders of reality to lower ones. God, who in His Essence is totally beyond form, number, matter, energy, space and time, must—as Frithjof Schuon never tired of pointing out—'overflow' into these dimensions of existence simply because He is Infinite; no barrier exists in His Nature which would prevent the radiation of His super-abundant Being.

The face of the Divine turned in the direction of this overflow into manifestation is the First Intellect, the Logos. The Logos may be considered either as God in the role of Creator or as the universal act of this Creator. In any case, the primal vibration of the Logos manifests, first, the spiritual universe, then the psychic, then the subtle or animic, and lastly the material. The 'bursting through' of the creative impulse from the animic to the material plane is what physicists have named the Big Bang. This is the point at which God's creative act enters time, thus creating it. 'Before' this point, the 'stepping down' of creation to less ordered levels is expressed in terms of the ontological precedence of one plane of Being over another, not in terms of successive temporal stages. The temporal unfolding of the universe is nonetheless a reflection, on a lower ontological level, of the eternal Hierarchy of Being as it exists on higher levels. In the words of Plato from the *Timaeus*, 'Time is the moving image of Eternity.'

From the standpoint of the material level alone, it seems that the state of the universe at the Big Bang was materially simpler than the universe is today. Simple hydrogen atoms came later than the Big Bang; more complex atoms, later than that; molecules, still later; and the complex structures of life even later. How, then, can the Second Law of Thermodynamics, which scientists as a whole accept, be true? If material structures are becoming more complex, how can entropy be valid?

Entropy is defined as the inevitable decrease in temperature differentials between the various parts of the universe. The total amount of heat is the same, but higher temperatures become progressively rarer. Where the temperature is higher, the frequency of the electromagnetic energy is also higher—higher-frequency blue stars, for example, are hotter than red stars—and higher frequencies of energy can carry greater amounts of 'information', which is another way of defining 'order'. According to the laws of physics, the creation of complex structures of matter necessitates a greater expenditure of energy, a greater even-ing out of the overall temperature of the universe, and thus a greater increase in entropy, than the order which is gained by the increasing complexity of matter can overcome. So, although the complexification of matter, which reaches its greatest extent in life—specifically human life—is apparently an-entropic, the net entropy of the universe is still always increasing.

But is an increasing complexity of material structures really an-entropic? In terms of the net entropy of the universe, clearly it is not. The higher temperatures of the early universe, while they worked against the complexification of matter, apparently made possible, in the form of energy, a higher density of 'ordering information' than did the later and lower temperatures, which allowed matter to coagulate and complexify; this is the only conclusion possible in view of the fact that increase in entropy and increase in material complexity seem to go hand-in-hand. A higher degree of order must therefore be possible in certain states of energy than is possible in any state of matter. Such a higher state of order must be expressed in terms of a synthetic simplicity, as opposed to lower states, which must tend in the direction of analytic complexity. Therefore all material structures must be incomplete, lower-level translations of higher and more capacious levels of ordering energy, just as the material universe as a whole—matter, energy, space and time—is a lower-level translation of the eternal hierarchy of Being as it exists 'before' the Big Bang.

Therefore, in terms of the entire thermodynamic economy of the universe, more complex forms of matter—Man himself being the most complex—actually represent higher states of entropy, not in relation to less ordered states of contemporary matter, but in relation to earlier stages of the universe when more information was stored in energy and less in matter than is presently the case. (In still later stages of the universe, entropy may have increased to the point where complex material structures again become impossible, not because the density of ordering energy is greater than will allow 'room' for such structures, but because it is less than will support them.) However, the complexification of matter *is* an-entropic in another, and special, sense: insofar as it gives rise to the potentiality for self-reflexive consciousness, latent in simpler life-forms but fully-formed in Man.

But this an-entropic re-concentration of energy, this flow of existence back to its 'earlier' energy-levels through ascension to higher degrees of order, higher levels of Being, is not material.

To believe that it is material—to project upon the material dimension what can only happen in the dimension of consciousness—is the fundamental error of the New Age, as well as of all evolutionary-utopian conceptions of history.

According to the Qur'an, *Unto God all matters are returned.* But they do not return to Him materially. The universe does not again contract till it forms the 'primeval atom' which supposedly came before the Big Bang. Materially, it expands, its entropy increases, until it approaches—but never reaches, at least in material terms—the pole of Substance, the *prima materia*, which in Aristotelian philosophy (especially in its esoteric implications as developed within Islam) is the pure formless receptivity underlying all matter—the 'waters' upon which the Spirit of God moved in *Genesis.* The universe returns to its Source in only one way: through the consciousness of self-reflexive beings who, by virtue of the free will inherent in this self-reflexiveness, have chosen to transcend themselves, to return consciously to their Creator, to walk the Spiritual Path. The macrocosm can only expand and decay; the conscious microcosm alone, insofar as it dis-identifies with this expansion and decay, becomes the avenue for the an-entropic flow of all things back to God. Only Man, and other self-reflexive and potentially self-transcending beings in the universe (since according to traditional metaphysics the Human Form is the synthesis of all material and immaterial creation) can bring the universe back to the 'primeval atom'. This is precisely the 'gathering of the scattered sparks of the Godhead' which constitutes the *tikkun* or universal restoration in the Kabbalah of Issac Luria. In Aristotelian terms, it is Man's return to the pole of Essence, the *imago dei* which dwells within the spiritual Heart, by virtue of his ascent along the ontological ladder of his own being. This Essence is expressed in terms of the Logos: the ordering energy of all levels of universal manifestation at its greatest point of synthetic simplicity. It is the Image of God, eternally radiating its own Being, through its Logos, its Word, into dimensional existence. The echo of this eternal radiation, in the world of space and time, is the Big Bang.

Manifestation is by nature expansive and entropic. If there

were no entropy in stars, they would not shine, and consequently nothing could be seen. If there were no entropy in matter, then the friction which produces sounds and tactile
sensations would not exist; the universe would be silent, and
numb. Without entropy, the processes by which conscious life
is materially embodied would not exist, nor would any potentially embodied life be capable of experiencing its environment,
including its spacio-temporal self, since no information from
this 'environment' could ever reach it. And there is absolutely
no logical way of determining whether the environment experienced or the capability of experiencing it has precedence, since
neither is possible or conceivable without the other: where
there is no existence, there can be no experience; where there is
no experience, there can be no existence. This is the real significance of the 'anthropic principle', which is essentially the same
as the doctrine of the Primordial Man, found throughout traditional metaphysics and mythology. According to this universal
doctrine, God created the universe first in the form of the Primordial Adam, the *insan i-kamil*—the eternal self-reflexiveness
inherent in the Divine Nature—who contained within himself
the entirety of spiritual and material creation. The manifestation of the Primordial Adam, the 'motion' from eternity to time,
is visible. The reintegration of the manifest universe as the form
of the Primordial Adam, the 'motion' from time to eternity, is
secret and invisible. If the Big Bang radiates energy, the Spiritual
Path, as it were, absorbs it; the Path is, in the words of Seyyed
Hossein Nasr, 'the reversal of the cosmogonic process.' (According to Stephen Hawking in *A Brief History of Time*, even black
holes do not absorb energy such as to violate or reverse the Second Law of Thermodynamics.) This is the true meaning of the
Hindu concept of a cyclical universe, termed 'the outbreathing
and inbreathing of Brahman.' In Muslim terms, it is the polarity
between *ar-Rahman*, God's all-manifesting creative mercy which
generates the macrocosm, and *ar-Rahim*, His particular and
saving mercy which, as the principle of the Spiritual Path, is
ultimately directed only to the individual, the microcosm.
The analogous concepts from Hinduism are *Avidya-Maya*, the

manifestation of God which progressively veils Him, and *Vidya-Maya*, the intimation of God as transcending manifestation, which ultimately reveals Him. In the words of St. Paul from First Corinthians, 'It is sown a physical body, it is raised a spiritual body' (I Cor. 15:42).

The psycho-material universe, then, might be termed a 'subject/object wave', where objective, material existence is constantly radiating, expanding, and increasing in entropy, while subjective, psychological existence—insofar as it truly transcends itself, and thus realizes the indwelling Subject, the *atman*, the Absolute Witness which is higher than the psyche—is eternally gathering and concentrating itself an-entropically, eternally rising to higher levels of order and Being, which are higher levels of not of analytic complexity, but of synthetic simplicity.

The objective, sensual universe manifests in terms of energy, the subjective universe in terms of consciousness. *Experience* is the process whereby energy is transformed into consciousness. If experience is not also transcended, however, the consciousness it produces remains on the level of memory, which is merely psychic. The memory of physical events, stored on the psychic plane (in what the Hindus call the 'akashic record') in the form of 'impressions' (*sanskaras*), becomes the seed of future events; this is the process of karmic causality resulting in 'reincarnation', though in reality no single individual soul reincarnates; what is transmitted is merely the potentiality of event/experience, via lesser subject-object waves contained within the single great subject/object wave which is the psycho-physical universe. The full return of material manifestation to its unmanifest Source via the road of consciousness only happens in those cases of 'annihilation in God' or 'union with the Godhead' or 'perfect total enlightenment', where the psychic plane—experience itself—is transcended.

The subject-object wave we call the universe is, like all waves, a cycle—which is why our word 'universe', in Latin, means 'one turn'. In terms of space and time, the material universe seems to give evidence of its own cyclical nature; such evidence is never perfect, however, and always contradicted by other evidence, because the universe, as a subject-object wave, is in reality a *standing* wave, like the vibration of a plucked string. Unlike a wave of the sea, it does not 'pass' in time (even though it is 'composed' of time) since there is no time-frame outside it through which it could pass. This quality of time, greater than passing time but less than Eternity, is that denoted by the Greek word *aion;* it is what Eastern Orthodox Christians call 'aeonian time'. If there is only one cycle, then the material universe, depending upon our point-of-view, appears either as cyclical or as eternal, precisely because, as a product of the first 'event' in spacio-temporal manifestation, it essentially exists on the border between time and Eternity.

Scientists can posit a Big Bang, and find evidence of it through its effects. What they admittedly cannot do is answer the question: 'Why was there a Big Bang in the first place?' Reality 'before' the Big Bang is also 'before' (in metaphysics, we would say 'ontologically prior to') matter, energy, space and time. Therefore none of the physical laws which describe the behavior of matter, energy, space and time can explain why, or if, the Big Bang was necessary. It might just as well never have happened.

The fact that the Big Bang, which produced all material manifestation, did in fact happen, coupled with the impossibility of declaring that such an event was necessary, is the reflection, in physics, of the principle, from metaphysics, of the Transcendence and Immanence of God.

God, according to traditional metaphysics, is necessarily both absolutely beyond the universe—since He Himself is the Absolute while all manifestation partakes of relativity—and also entirely immanent within it, pervading it everywhere, everywhen, and in every mode. Because God is Absolute and Infinite, He cannot be contained by the dimensional, quantifiable universe. But—because He is Absolute and Infinite—He necessarily also pervades it; it's as if the universe were actually contained in Him.

The principle of God's Transcendence and Immanence means that the question 'is the universe God?' always generates two answers: Yes and No. The material manifestation of 'yes' is the Big Bang. The quasi-material manifestation of 'no' is the impossibility of proving, on the material level, that the Big Bang was necessary.

This 'yes' and 'no', eternally co-existing, are the origin of the eternal vibration of the Logos, the plucking of the first and only string on God's eternal lyre, whose result, on the material level, is the standing wave we call the universe. The eternal answer 'yes' is the peak of the wave made by that vibrating string; the eternal answer 'no' is the trough of the same wave. Thus manifest Being is woven of existence and non-existence—and the vibration between the poles of existence and non-existence is the Logos itself, the primordial musical tone, the syllable Om, the first Word spoken, eternally, by God the Creator, when He created a universe which was, by metaphysical necessity, both Him and not-Him. This is one meaning of the first verse of the Gospel according to John: 'In the beginning was the Word, and the Word was *with* God, and the Word *was* God.' From the standpoint of the Logos, the manifestation and re-integration of the universe— creation and *apocatastasis*—are simultaneous, not successive.

Traditional metaphysics represents this quality of manifestation as composed of existence and non-existence by such symbols as the T'ai-chi or yin-yang sign; by the Hindu doctrine that all manifestation is *maya*, the magical self-revelation of Brahman, something which is never entirely unreal but never quite what it seems; or by the Buddhist concept that all manifestation is intrinsically void: in the words of the *Heart Sutra*, 'Form is emptiness, emptiness is form.'

Physics reflects this same concept, on a lower level, in such doctrines as that of matter and anti-matter, where 'empty' space emanates a particle, leaving a 'hole' in this space as the corresponding anti-particle.

The understanding that universal manifestation both exists and does not exist, because it is both God and other-than-God, is the basis of the Spiritual Path: the an-entropic return, via the spiritual Heart in self-reflexive beings, of energy—as experience both lived and transcended—to its Absolute Source. If the universe were God, such return would be unnecessary, since material conditions themselves would be Divine. If the universe were not God, such return would be impossible, since no sign of God would exist to inform us of His Reality. But since the universe both is and is not God, it becomes the ladder of return to Him, a ladder with both rungs and the spaces between them. In order to climb this ladder, the reality of our experience of the world must be both affirmed and denied, both embraced and transcended. In traditional mystical terminology, the affirmation is called *cataphasis*, and the denial, *apophasis*. In the true practice of contemplation this affirmation, and this denial, happen with every conscious breath.

cV: The Slowing of Light

Physicist Barry Setterfield, among others, has noted the apparent fact, based on measurements taken since the 1700s, that the speed of light is slowing down; in other words, *c* is not really a constant. I would assert, however, that there is no way to prove this contention objectively; all that could be proved would be that the ratio between the speed of light and the 'clocks' we use to measure it is changing; there is no third, strictly objective 'Archimedian point' outside these two criteria which we could use to determine whether light is slowing down, or whether in fact the physical processes we use to measure light's velocity are speeding up. In order to measure the speed of light, we need to assume that the physical process against which we measure it—the vibration of a quartz crystal or a cesium atom, for example—is constant. But if the 'grand constant', the speed of light, could vary, who is to say that the rate of vibration of a crystal could not also vary?

If I run the hundred-yard dash on two successive days ('I' am light in this analogy), and the clock that times me doubles in speed from one day to the next, it will appear that my speed on Wednesday is only half of what it was on Tuesday. And if the clock that times me is the only clock there is, then there is absolutely no way to tell if the clock has sped up or if my running speed has slowed down. The case could even be made, thinking relativistically, that the question 'Am I slowing down or is the clock speeding up?' is meaningless. All that can be determined for sure is that the ratio between the two speeds is changing. And is it intrinsically any more likely that light should be slowing down rather than that all physical processes exhibiting periodic motion should be speeding up? If we were to operate under the assumption that the 'clocks' we use to determine the speed of light were in fact speeding up, then the measurements Barry Setterfield has have used to present evidence of the fact that light is slowing down would not change one iota; only their interpretation would change.

French metaphysician and mathematician René Guénon, in his book *The Reign of Quantity and the Signs of the Times* (1945), follows for the most part the Hindu doctrine of cyclical time, in which earlier world-ages or 'yugas' exhibit more 'spacial' characteristics, while temporality becomes increasingly dominant in later ages as the cycle descends toward its conclusion. He says this:

> .... according to the different phases of the cycle, sequences of events comparable to one another do not occupy quantitatively equal durations; this is particularly evident in the case of the great cycles, applicable to both the cosmic and to the human orders, the most notable example being furnished by the decreasing lengths of the respective durations of the four *Yugas* which together make up a *Manvantara*. For this very reason, events are being unfolded nowadays with a speed unexampled in the earlier ages, and this speed goes on increasing and will continue to increase up to the end of the cycle; there is thus something like a progressive 'contraction' of duration....The increase in the speed of events, as the end of the cycle draws near, can be compared to the acceleration which takes place in the fall of heavy bodies.

Barry Setterfield's findings would seem to corroborate this assertion.

Guénon also spoke of the 'solidification of the cosmic environment' as a symptom of the approaching end of this cycle of manifestation or *manvantara* (for *manvantara* read *aion*). Given that light slows down when passing through a dense medium, the slowing of the speed of light could also be read as an index of the densification of the cosmic matrix, which Guénon also defines as a sort of increasing 'impermeability to Grace'.

I believe that Einstein posited the speed of light as a constant in order to make *c* stand for the presence of God in the universe. For all his 'relativity', he still maintained an intuition of God as an absolute objective Fact, an intuition that also made him uncomfortable with quantum mechanics and its principle of indeterminacy, which can only be approached through probability, not strict determinism; this is why he said 'God does not play dice with the cosmos.' But God is not measurable; He is absolutely transcendent. 'None has seen the Father at any time.' To replace God with a quasi-absolute measurable constant, a physical quantity, is essentially idolatry.

Yet light still remains a potent *metaphor* for God. In the words of the Qur'an, *God is the light of the heavens and the earth.* (The Qur'an also makes clear that light is not to be worshipped as God, in its story that Abraham first worshipped the stars, then the moon, then the sun, but they all set—after which he declared 'I will no longer worship things that set,' indicating that he had realized that God is eternal, and is not one of the objects of the universe.) According to relativity theory, as an observer speeds up, his or her time slows down in relation to his surrounding matrix. If the observer were able to reach the speed of light, his time would stop, and the speed of time in his surrounding matrix would become infinite. Furthermore, relativity theory maintains that whatever increases in velocity also increases in mass; therefore the mass of an 'object' traveling at the speed of light would be also infinite. And though such a condition—in which the 'observer' does not see light but rather *is* light, while also possessing infinite mass, can be reached only theoretically, this condition is precisely Einstein's metaphor for eternity (the transcendence of time) and infinity (the transcendence of spacial and material limits).

Such a state might represent the way things were 'before' the Big Bang—except for the fact that eternity is not essentially 'before' or 'after' time. Time cannot really begin or end 'in' time, since for this to happen there would have to be a time 'outside' time and encompassing time, and this is a contradiction in terms. In reality, it is eternity which encompasses time, embracing its beginning and end in a single cycle: 'Before Abraham came to be, I am.' This is precisely the meaning of the Greek word *aion*.

Eastern Orthodox theology has the concept of *aeonian time*, which is not linear, uni-directional time but cyclical time: a period or cycle of time conceived of as a single eternal form. 'Behold I am with you always, even to the end of the *aion*.'

To *be* light, to exist in timelessness, to be beyond subject and object (since if the 'observer's' time stops and the time of the 'observed' becomes infinite, then the terms 'observer' and 'observed' become meaningless), and to possess infinite mass—this being a material metaphor for infinite Reality, infinite Being—is to exist as we did before the Fall (or, dare I say, before the Creation). The 'fall' of humanity away from this original Unity was, in Biblical terms, progressive: the eating of the fruit was the first Fall, the slaughter of Abel a further fall, the Flood another, the change from rule by prophets to rule by kings another, the change from Tabernacle to Temple another, the exile another, the destruction of the second Temple still another, etc., etc. In terms of physics, the same progressive descent might be evidenced by the slowing of the speed of light and/or the speeding up of time, which—mysteriously—is very close to Einstein's view of what would happen to an object, vis-a-vis its environment, if it were to approach the speed of light. While we remain *observers*, identified via our act of observation (which is an act of *separation*) with the material universe in which we live, we continue to fall. The slowing of light is an apt metaphor for our progressive alienation from Divine Intelligence, which is related to the *intellection* of the scholastic philosophers: a form of knowing which is both instantaneous and complete. As the flow of events (time) speeds up, the 'rate' of intelligence—which requires a deep, contemplative, timeless calm, a *spacial* quality, in order to develop to its full capacity—'slows down'. And this is precisely what is happening everywhere in human society. The slowing of light, the speeding up of time, is dumbing us down, dumbing *everything* down. It is not simply a case of 'social breakdown', but an actual deterioration of the cosmic environment.

So we are falling away from God into matter, from eternity ever further into time—*if*, that is, we retain our self-identification as 'observers'. But if we transcend, through spiritual illumination, the subject/object dichotomy—if we no longer, by positing ourselves as separate observers, egoistically oppose ourselves to God's will as manifest in the changes in the cosmic environment at the end of the cycle, in the 'latter days', then—speaking metaphorically—the speed of time will become 'infinite' and the speed of light will be 'zero'. *O felix culpa!* (The *Zohar* expresses this state by saying that 'In the messianic age, the object and its image will no longer be related.') And, I would assert, there is no essential difference between the statement that the time of the 'observer' (identified here with physical processes exhibiting periodic motion) has become infinite and the time of the 'observed' (identified here with light) has been reduced to zero, and the reverse statement, that the speed of light is infinite and the speed of physical processes, zero. In both cases the condition known as *eternity* has been 'reached': post-eternity and pre-eternity are equally *eternity*. And since the distinction between subject and object, observer and observed has been annihilated in both cases, I would submit that we are no longer dealing with two cases, but only one. It is as if, in the annihilation of the distinction between matter and light, light has become matter and matter, light. Such a condition is certainly not physically or materially attainable, since its definition transcends all perceptible and measurable limits, all material definition. But it remains as valid a metaphor as any for the relationship between eternity and time, and for the precedence of eternity over time—at the 'beginning', at the 'end', and virtually also in the *now*, since eternity is a Now which encompasses time but cannot be encompassed by it: 'I am Alpha and Omega, the beginning and the end; before Abraham came to be, I am.'

In *The Reign of Quantity and the Signs of the Times*, Guénon expresses this 'consummation of the age' in the following terms:

In an earlier chapter it was stated that in a certain sense time consumes space, and that it does so in consequence of the power of contraction contained in it, which tends continuously to reduce the spacial dimension to which it is opposed: but time, in its active opposition to the antagonistic principle, unfolds itself with ever-growing speed, for it is far from being homogeneous, as people who consider it wholly from a quantitative point of view imagine, but on the contrary it is 'qualified' at every moment in a different way by the cyclical conditions of the manifestation to which it belongs. The acceleration of time is becoming more apparent than ever in our day, because it becomes exaggerated in the final periods of a cycle, nevertheless it actually goes on constantly from the beginning of the cycle to the end: it can therefore be said not only that time compresses space, but also that time itself is subject to a progressive contraction, appearing in the proportionate shortening of the four *Yugas*, with all that this implies, not excepting the corresponding diminution in the length of human life [as measured against the lifespans of patriarchs like Methuselah, Guénon means—which denote, in my opinion, not 'more years' but more *actual time*, a greater richness of experience, per year, per day, per second]. It is sometimes said, doubtless without any understanding of the real reason, that today men live faster than in the past, and this is literally true. . . .

If carried to an extreme limit the contraction of time would in the end reduce it to a single instant, and then duration would really have ceased to exist, for it is evident that there can no longer be any succession within the instant. Thus it is that 'time the devourer ends by devouring itself,' in such a way that, at the 'end of the world', that is to say at the extreme limit of cyclical manifestation, 'there will be no more time'; this is also why it is said that 'death is the last being to die', for wherever there is no succession of any kind, death is no longer possible. As soon as succession has come to an end, or, in symbolic terms, 'the wheel has ceased to turn', all that exists cannot but be in perfect simultaneity; and this can also be expressed by saying that 'time has changed into space'. Thus a 'reversal' takes place at the last, to the disadvantage of time and the advantage of space: at the very moment when time seemed on the point of finally devouring space, space in its turn absorbs time; and this, in terms of the cosmological meaning of the Biblical symbolism, can be said to be the final revenge of Abel on Cain.

Jesus passed through this 'pole shift', this condition where time devours space until it suddenly 'becomes' space, in His crucifixion and resurrection. In other words, He went through on the cross what for the rest of us will be the Apocalypse. Yet, as William Blake said, 'when a man rejects error and embraces Truth, a Final Judgement passes upon that man.' Speaking in Christian terms, if we 'die with Christ' and 'rise with Him', our spiritual center has entered and passed beyond the Apocalypse already. Time, in devouring itself, has become 'space' (Guénon's metaphor for eternity); eternity is realized. Consequently, whatever may happen to us externally and materially in these apocalyptic times, our essence, at rest in God, will not be touched by the 'second death'.

PART FOUR

Sparks from the Anvil

On Reincarnation

The whole question, when it comes to reincarnation, is: who or what reincarnates? In the theory of reincarnation as it is usually taught in the west, three things are confused: reincarnation, transmigration and metempsychosis. Transmigration has to do with the journey of the human soul not from one physical human life to another, but to higher (or lower) planes of being; as Jesus said, 'in my Father's house are many mansions.' Metempsychosis is the inheritance of psychic material. Just as a person who dies will leave clothes and other possessions behind which can pass to others, so the deceased can also leave behind psychic 'residues'—even memories. When we inherit such memories, we may mistakenly believe that we actually lived before as other persons in past lives. Such memories or psychic 'substances' are usually inherited when the soul destined for human life is born into this world; but they can also be inherited at any point during our earthly life. When someone leaves this world, many such residues may be left behind (especially when the person in question was not particularly integrated psychically)—but such residues can also be liberated when someone undergoes any radical life change, residues which can then be picked up by others—'ghosts of his former self'. For example, I just moved to Kentucky after a lifetime in California—and a man I know seems to be picking up what I left behind there. He lives in New York State but has fallen in love with a woman from my home town; he keeps traveling to California as if retracing my steps, ending up visiting all the places I once inhabited, even trying to live out some of my former ambitions—all of this unconsciously. It's as if my 'former incarnation' (myself in California) has 'reincarnated' as him, or as if his 'former self' has died and 'reincarnated' as my 'former self'. That's metempsychosis. It happens entirely on the psychic plane, while transmigration is travel of the true self through the Spiritual world (which is higher and more inclusive than the psychic world), through the different planes of Being. And reincarnation, as usually taught, is simply a fallacy; I, in my full and unique humanity, cannot 'turn into' somebody else. I can transcend myself, die to myself, realize my essential identity with the Absolute, but there is no way (thank God!) I can become someone *other* than myself. In any case, insofar as eastern religions teach reincarnation (if indeed they do), they do not usually portray it as a method of spiritual growth and evolution, but more often as a hell of perpetual incompletion, in which we are always *becoming* but never quite able to BE. It is something to be liberated from, not something to 'use' or 'explore'.

We *want* to believe in reincarnation only because our intuition of eternal realities is obscured. If we can't imagine non-incarnate states of being and consciousness, then the only way we can conceive of Eternal Life is as an endless series of 'other' lives in *this material world*, or other worlds like it. Reincarnation is essentially a materialistic belief.

You speak of your experience of 'past-life regression—but how can you possibly be sure what you are experiencing in those regression sessions? The psychic world endlessly changes and mutates; there is no stability there, and therefore no certainty. The *only* way to reach any certainty about the state of one's psyche is to vow to transcend it—after which all its ambiguous tendencies will fall into two categories: those which submit to God's command that we transcend ourselves, and those which oppose it.

On Changelings

Folklore speaks of what are called 'changelings'. The belief was that the Fairies would steal a human infant and leave a fairy child in its place. This child looked exactly the same as the human one, yet was in some way uncanny. This probably has to do with a half-conscious insight on the part of adults into the nature of the unregenerate *nafs-al-ammara*, the 'natural' soul which seems so innocent in the infant, but which will be the origin of all evil later in life unless it is subdued, refined and purified. William Blake, in reaction against the sentimental and unrealistic idealization of the state of infancy, wrote:

> *My mother groan'd, my father wept*
> *Into the dangerous world I leapt*
> *Helpless, naked, piping loud*
> *Like a fiend hid in a cloud*

Christian culture understood that the uncanny, dangerous and unregenerate soul of the infant was under the power of 'original sin'—the effects of the Fall—until purified and protected from evil by the sacrament of baptism. But in post-Christian culture we idealize the infant, and our own infantilism (Michael Jackson!), and consequently we are without the kind protection from the Jinn that our ancestors enjoyed.

Folklore also tells of marriages with Fairies. A man falls in love with a woman of the fairy realm and she agrees to marry him, on condition that he does not break this or that taboo. The marriage takes place, but of course he does break that taboo later, usually through curiosity, and his fairy wife flies into a rage and leaves him forever. These are most likely 'imaginal' marriages, like those with a demonic succubus or incubus or familiar spirit—though the possibility of a spiritually positive liaison of this sort appears in the relationship between a Vajrayana adept and his *dakini*, his spiritually helpful female consort on the Imaginal Plane.

In our time, however, we can see the effects of a progressive 'jinnifying' of the human psyche. The fascination with psychic powers and the increasing permeability of the energy-barrier between the material and intermediate planes that René Guénon, in *The Reign of Quantity*, saw as opening our world to invasion by the 'infra-psychic', has produced a kind of volatilization of the soul. Certain psychic functions, such as intuition and fantasy, are becoming demonically swift and penetrating, leaving others—logical thought, for example, or normal human empathy—depleted and petrified. In the words of the Qur'an, humanity in the latter days shall become *like scattered moths*. And this elvish 'changeling' quality can affect our bodies as well, as in the martial arts and X-treme Sports movements, for example. Some people are almost learning to fly in their physical bodies, while simultaneously playing in an uncanny and cavalier manner with mutilation and death. Such people show every symptom of 'pixilation': possession by the Powers of the Air. This is simply one more attack on the integrity of the human form. While globalism and technocratic totalitarianism mercilessly attack from without, the temptation to abandon our humanity and cast our lot with the Fairy Folk as refuge from them works insidiously from within. In the words of W.B. Yeats:

> *Come away, O human child*
> *To the waters and the wild*
> *With a fairy, hand in hand;*
> *For the world's more full of weeping*
> *Than you can understand.*

But to abandon the human form, or adulterate it with elvish glamour, is to betray our only possible effective relationship to our Creator. In the words of Jesus, 'None come to the Father but through me.'

On the Manipulation
of Psychic Energies

Taoism, Shinto, and the Vajrayana all stem from the genius of the Mongolian race, whose primordial spirituality was shamanism. In world ages when the level of Spirit is generally understood and not heavily veiled, it is possible to access and use psychic and psycho-physical forces by the power of the Spirit and under Its command. In a few instances, this may still be possible today. But in the vast majority of cases, the manipulation of psychic powers does not transcend the psychic plane, and so becomes vulnerable to infiltration and domination by the infrapsychic. (In C.S. Lewis' fantasy novel *That Hideous Strength*, Merlin returns from 'occultation' to ultimately aid humanity in its struggle against forces which are effectively those of the Antichrist; but it is made clear that his theurgic powers, his nature-magic, are ambiguous and in many ways anachronistic in our times, and therefore very dangerous.) Ch'i or Ki (the Taoist equivalent of the Hindu *prana* or life-energy), in its highest sense, is like the *shakti* of the Spirit; when Spirit is fully realized, ch'i (one of whose manifestations is the kundalini) becomes its direct expression. But when the level of Spirit is veiled, ch'i becomes ambiguous. Conceived in lower terms, it then operates on a lower level.

On Psychedelics

Though the Native American Church seems to provide a better quality of life to Indians on the Res than alcoholism does (which is not saying much), history shows that the mass use of psychedelics (I'll continue to call them that) did not lead to 'a more loving culture'; quite the reverse. What they did was rupture the natural psychic protection of thousands or even millions of people, making them too 'psychic', or at least too 'psychically empathetic' in an imbalanced way. In other words, psychedelics created radical 'boundary problems'. When people are too sensitive to the inner feelings and psychological experience of other people, it does not always bring them closer to others, but often drives them further apart. Over-sensitive people may become paranoid, or freeze up emotionally, or become reclusive or agoraphobic. The presence of others is often painful for them, causing them to flee human contact, or to relate to others in a shallow manner, either emotionally wooden or sometimes scattered and manic. The prerequisites for healthy relatedness are self-containment, recollection, centeredness, discretion. If you lose yourself, or if you identify so totally with someone else that you almost feel you *are* them, then you cannot really relate to them. To relate to others you have to respect them, and you can't respect them without respecting yourself. To 'respect' means 'to look again'. Instead of just flowing into immediate empathy and identification with other people, we need to pause and 'look again'; we need to intuit the true psychic boundary between ourselves and others—it will be fluid, but it will always be there, a true objective factor. This is the basis of courtesy. Without *courtesy* we will unconsciously identify with others, which means that we will treat them as if they were parts of ourselves. In a worse case scenario, we will 'possess' the other while simultaneously being possessed by him (or her); this is the final end of what is called 'co-dependency'. If I am stuck inside you and you are lost inside me, then we can never meet. And all of these are problems which psychedelics can seriously exacerbate.

A further problem is the use of psychedelics by some people in pursuit of psychic powers. Developing psychic abilities may give one a temporary sense of relief from constriction; I'm reminded of a prison convict I once corresponded with who was practicing astral projection—it's not hard to imagine why. But it is not a healthy course to pursue, and if one is serious about the spiritual Path, then it is most often a fatal one. This was a common enough idea to be a cliche in the hippy days when we were studying eastern religions: pursuit of psychic powers delays or totally blocks spiritual progress. (If only we had taken our own advice!) When you are addicted to psychic experience, it becomes nearly impossible to fix your attention on the One. The goal of the spiritual Path is to stop being centered in the psyche and to establish, as one's Center, the Spirit. It is to stop self-identification—but psychic experience IS self-identification; oneself as the experiencer is always the main protagonist. (I was always going to write a book entitled *How To Lose Your Psychic Powers in Ten Easy Lessons*—Lesson One being *Mind Your Own Business*.) God may give you a psychic power for His own purposes—to put you through needed experiences or to help you serve—but to *seek* psychic powers is frowned upon by most traditional authorities. Shamanism is the big exception, and the only spiritual justification for shamanism is service to one's tribe: healing, finding game, bringing rain etc. But self-aggrandizement and various forms of ego-based sorcery are always big temptations in the shamanic world; even Black Elk—who was actually a Catholic catechist among the Lakota for longer than he was a traditional medicine man—said that many of the medicine people he knew were dangerously arrogant. The Siberian shamans say that 'God sent shamans to the earth to fight demons,' and recognized that the strain of this warfare often shortened their lives.

On Tolkien

J.R.R. Tolkien is seen by traditional Christians as a writer of Christian allegories—the 'King' who is to 'return' is Christ, etc.; and this is in fact the case. But hardly anybody else saw him that way, and it is certainly true that his *Lord of the Rings* trilogy was taken by actual or incipient Neo-Pagans as paganism pure and simple; even before the movies came out, it represented a great stride toward the spurious popular conception of a non-Christian Middle Ages. Look at Hollywood films about King Arthur from the 50s and early 60s; there was no question but that he was a Christian monarch. People like Ezra Pound, of course, hated Christianity, and this led them to an interest in the more heterodox aspects of the medieval ethos, such as the tradition of Courtly Love. (Courtly Love, to my mind, is still essentially Christian—as we can see from Dante—or at least impossible in that particular form outside a Christian context.) But now, after Tolkien and a lot of other fantasy writers have done their work, the popular concept of the Middle Ages, or something like them, is now purely one of warriors, demons and wizards, with hardly any Christian coloring at all. As western culture becomes post-Christian and Neo-Pagan, it has to rewrite its own history.

The Lord of the Rings in itself expresses a valid level of spiritual truth; it is like the Final Judgment, or the days leading up to it, as seen from the standpoint of the Imaginal Realm. But I don't believe it can really help us in confronting these apocalyptic times in the specifically *human* way that is required of us—though we certainly need to be reminded that power is not to be held by the little egoistic self (this being the Titanic error) but must be sacrificed to a greater Reality; I'm referring here to Frodo's ultimate sacrifice of the One Ring.

Partly under Tolkien's influence, some in my generation seriously attempted to conclude an alliance between the human race and the Nature Spirits, a kind of united front against the technocratic culture that was, and is, destroying the earth. But such an alliance is no longer possible, if it ever was; the Elves have indeed departed from Middle Earth in their fairy ships; they can no longer aid us. (I am reminded of one night when, in the period years ago when I was running after the Nature Spirits, my wife and I, walking through a coastal forest in California—and without benefit of psychedelics—believed we had witnessed an elf's funeral.) California hippy culture took Tolkien's books terribly seriously. With the help of psychedelic drugs, and the beautiful forests surrounding us, some of us thought we could actually live in worlds like that. I know of people in the town of Bolinas, California who *literally* tried to live as Hobbits. They almost succeeded. It was very sad. And the thinly-veiled reference to marijuana at the beginning of the motion picture version of *The Fellowship of the Ring* proves that the tendency of neo-hippy or Neo-Pagan culture to claim *The Lord of the Rings* as its own is still in full force.

The closest thing to an actualization of the belief that we could ally with the Elves to combat the forces of materialism was the Findhorn community and its myth. It seemed to many as if that alliance had almost actually taken place. Those huge cauliflowers were undeniably real....but then so many who had opened themselves to the world of the Nature Spirits lost their hold on the human form through too much fairying-around; they became *pixilated* ('pixie-led'). And without that firm hold on our own essential reality, whose central seal for the Christian world is Christ's incarnation—without faithfulness to what the Qur'an calls the Trust, to humanity's pivotal role as God's vice-regent in earthly reality—everything fell away into alienated fantasy, into the attenuation and diminishment of the human form. And into that spiritual vacuum rushed all the demons that beset us today, the ghosts, the devils and the space aliens.

There's some of the secret, inner history of the counterculture, told partly in Tolkien's manner, in terms of the chronicles of the Imaginal Realm. The lesson we learned through our psychic excursions into the worlds of the Nature Spirits—if indeed we learned it—was that the Elves cannot help us, and that our only way of helping them is to remain true to the human trust. In the Islamic world, what France and England call the Fairies are called the Jinn—and according to the Holy Qur'an, even though not all the Jinn are demons, some of them being pious Muslims, to worship the Jinn is actually to corrupt them, to divert both them and ourselves from the duty of attending—in our radically different and essentially incompatible ways—to the will and reality of God. There was even a Muslim saint who, when he saw that some of the Jinn were praying along side him, politely asked them to carry on their devotions somewhere else—because even the pious Jinn can be a terrible distraction. And given the massive resurgence of overt Paganism in post-Christian Europe, the works of Tolkien, though of great value as a mythopoetic rendition of the 'unseen warfare' of the Spirit of Truth against the forces of evil and illusion, are in danger of reinforcing rather than combating the transformation of our collective image of the 'other world' from the Heaven of God the Father to the Intermediary Plane of the Elves, the Fairies and the Jinn. We can no longer it take for granted, in post-Christian culture, that people will automatically understand a Christian allegory in Christian terms, especially when the name of Christ is never mentioned. Christians, in other words, cannot simply claim *The Lord of the Rings*; they must work to *re*-claim it.

On Syncretism

WHEN CONSIDERING the phenomenon of syncretism, we need to make a distinction between substance (the revelation itself) and accident (the already-existing materials utilized by the revelation). The Qur'an was revealed by God, through Gabriel, to Muhammad, peace and blessing be upon him. This was a direct vertical intervention of the Divine into human affairs, and the acts of God are not subject to 'historical influences'. However, this revelation necessarily made use of the materials available in this world in order (so to speak) to construct a body for itself. It drew upon an already-existing language, Arabic, as well as already-existing historical lore. But Islam was not deliberately 'stitched together' out of these materials in the manner of today's syncretists and their often 'globalist' agendas; the magnetic power of the revelation attracted the materials it needed, and rejected those that were incompatible with it. Thus we could say that the revelation was the host, and the history, lore, doctrines and practices drawn from earlier traditions were the guests. These were thoroughly Islamicized, 'converted to Islam' if you will. In Aristotelian terms, the pre-existing elements were *materia* and the Qur'anic revelation *forma*. When those Eastern Orthodox Christians influenced by modern 'orthodox' (heretical) theology, if not by their Protestant background, try to expunge Plato from Orthodoxy, for fear that otherwise Orthodox Christianity will somehow be 'half pagan', my answer has been that Plato was one of the greatest 'converts' to Christianity—as, in a Muslim context, he was also a 'convert' to Islam. Looking at parallels between different traditions on a shallow level leads either to a sense that all are *fragments* of the One Truth, a Truth which can only be fully revealed if the traditions are syncretized—either that, or to an absurd, postmodern belief in 'multiple absolutes', which results in cynicism and nihilism. But a deep sense of these parallels may take one out of the realm of 'comparative religion' entirely, via the shock of the realization that we are not encountering 'belief-systems', but a tremendous and objective Reality which any true revealed tradition, if it is still alive, can put us into effective contact with. In other words, the realization that all living traditions, from their different perspectives, are dealing with the same Truth can put an end to our wandering from religion to religion. If all airplanes are headed for the same airport, why not stay in our seat?

On Evil and Unreality

One of the clearest signs of a certain type of luciferian evil is the 'cavalier' attitude its slaves adopt toward questions of truth; nine times out of ten, 'Olympian laughter' is nothing but lurid, demonic chuckling. Evil, metaphysically speaking, is a negation, a depletion of reality or the sense of reality, a *privatio boni*. It is fundamentally absurd, and it acts to deplete our own sense of reality by tempting us to try and make sense of it on its own level; this is one of the ways it hooks us. A Satanist's lack of 'seriousness' with regard to his beliefs, his apparent playfulness in the face of evil (one is reminded here of Anton LeVey's attitude) is in no way a sign of his harmlessness; quite the reverse. Only the Good and the True is fully worthy, and capable, of being taken seriously—and to treat it with an ugly levity rather than the beautiful seriousness it demands is nothing less than an act of sacrilege.

Wit is worthy of the Good and the True because the Good and the True is worthy of defense against the slander of this world— and because the manifestation of the Absolute in terms of the relative always has an ironic edge to it, which is a sign that we are nearing the border between the two realms. But though wit is a great thing, seriousness is greater; there is nothing is more beautiful or more moving than a deep seriousness in the presence of the One Thing that deserves it.

On Antichrist

Speculations on the final form that evil and error may take for this cycle of manifestation are all-too-common in these troubled times. Apocalyptic prophesies, the speculations of science fiction, and the projections of social scientists and environmentalists place our lives against a background of all-pervasive and permanent anxiety. And a convenient figure for the constellation of this anxiety, particularly in the worlds of Christianity and Islam, is the figure known as *al-Dajjal*, or Antichrist.

It is said that although we infinitely far from God, God is infinitely near to us—nearer, according to the Holy Qur'an, than our jugular vein. From the standpoint of eternity, this is always true. But from the perspective of time—time in Plato's sense, as 'the moving image of eternity'—we can say that, even if we are not coming closer to God, He is coming closer to us. Our personal lives, as well as the life of collective humanity, are moving inexorably toward the moment of the great encounter, whether or not we are 'cooperating', whether or not we are 'traveling', whether or not we think we are 'ready'.

The ego always wants to appropriate all value to itself, especially those values—valid or spurious—which go by the name of 'spirituality'. The ego begins by claiming such values on the basis of pride; but when the dawning of God's material, intellectual and spiritual gifts threatens to be followed by the full sunlight of God's living presence, the pride of the ego is transformed into fear. (Fear was always the unconscious the root of that pride, but now this fear can no longer be avoided.) The moment has come when the ego knows, consciously or unconsciously, that it has to die. And so, in the face of imminent annihilation in the Absolute, it attempts to identify with and *become* the Absolute—not implicitly, as it always did, but openly and explicitly. In the name of this 'great crusade', it moves to press all the powers of the psyche into its service, under the battle-cry of *'Self-deification or Death!'*

When the ego explicitly attempts to deify itself in this manner, it becomes Antichrist. When the collective psyche makes the same attempt, on the field of global society, it generates the *regime* of Antichrist. This regime, since it is based on a counterfeit of the Absolute, necessarily expresses itself as false religion, incorporating the highest metaphysical principles the ego is capable of understanding—insofar as the ego, with its purely mental, psychic and passional approach, can be said to understand anything. (Those among the 'New Religions' which attempt to deify the passions—and even, in a few cases, entertain the hope of physical immortality—are clear signs of this.) It does this in hopes that, if it can gain possession of the eternal metaphysical principles, it can render itself immortal. Consequently, the dominant socio-political forces of the latter days can best be understood, in their most universal essence, as the ego's counterfeit versions of eternal metaphysical principles. This is why an understanding of traditional metaphysics and esoterism is vital to any meta-theory of contemporary social forces—not in order to manipulate such forces, this being the ego's usual approach to metaphysics on both the individual and the collective plane, a basically magical approach—but in order to avoid being conscripted by their counterfeits.

In my book *The System of Antichrist: Truth and Falsehood in Postmodernism and the New Age* (Sophia Perennis, 2001) I attempt to apply metaphysical principles to social forces in just this way. After the terrorist attack on the United States on September 11, 2001, I went back to *The System of Antichrist* to look for relevant passages. This is part of what I found:

> Globalism and One World Government, in my opinion, are not the system of Antichrist, though they are among the factors which will make that regime possible. I believe the system of Antichrist will emerge—is in fact emerging—out of the conflict between the New World Order and the spectrum of militant reactions against it. They are what the Bible and Qur'an call 'Gog and Magog'; they are part of the same system.

According to *Apocalypse* 20:7–8, '. . . .when the thousand years are expired [the millennium during which the devil is bound, identified by Orthodox theologians as the church age], Satan shall be loosed out of his prison, and shall go out to deceive the nations which are in the four quarters of the earth, Gog and Magog, to gather them together to battle: the number of whom is as the sand of the sea.' According to *The Apocalypse of St. John: An Orthodox Commentary* by Archbishop Averky of Jordanville, the meaning of *Gog* in Hebrew is 'a gathering' or 'one who gathers', and of *Magog* 'an exaltation' or 'one who exalts'. 'Exaltation' suggests to me the idea of transcendence as *opposed* to unity, 'gathering' the idea of unity as *opposed* to transcendence. The implication, here, is that one of the deepest deceptions of Antichrist in the last days of the cycle will be to set these two integral aspects of the Absolute in opposition to each other in the collective mind, and on a global scale, in 'the four quarters of the earth. . . .'

If all possible alternatives to the struggle between globalism and tribalism disappear from the collective mind, then Antichrist has won. He can use economic and political globalism and the universalism of a 'world fusion spirituality' to subvert and oppress all integral religions and religious cultures, forcing them to narrow their focus and violate the fullness of their own traditions in reaction against it. He can drive them to bigoted and terroristic excesses which will make them seem barbaric and outdated in the eyes of those wavering between a global and a tribal identification, and set them at each other's throats at the same time. Unite to oppress; divide and conquer.

If this is happening on the stage of global history, it is also happening, even more intensely and essentially, within the soul of each one of us. To recognize this is to come into the field of the great spiritual opportunity afforded by apocalyptic times. In *The System of Antichrist*, I say it like this:

> If you know your own ego, you know the Antichrist; if you know the God within you, you know God. The criteria by which we can recognize the Antichrist are the same as those by which we can recognize sin: If we understand what Divine Wisdom is, we will recognize what is contrary to that Wisdom; if we know what Divine Love is, we will be sensitive to what violates that Love. The signs of the end in the various traditional eschatologies cannot be

applied directly to history, without first being applied to the state of one's soul. Only after 'the discernment of spirits' is established within our own intellect, will, and affections can we turn and see the forces operating in these latter days of world history in the light of objective truth. If we know how the ego operates, especially when it attempts to appropriate our struggle against God in order to claim holiness for itself, or break its way into the mysteries of order to claim wisdom, then we will not be fooled by the analogous moves of the Antichrist on the field of history. . . .

To fear the Antichrist is actually to worship him. It is to fail to recognize in the ego's final attempt at self-deification a sign of its approaching end. Behind Antichrist stands the Messiah; behind ego shines the Presence of God. Our work is to attend, not to the distractions which veil that Presence, but to the Presence itself:

When I first saw the Antichrist, my response was: 'This means that I no longer have a single enemy on this earth. May all beings be well; may all beings be happy.' When Antichrist lived with me in my own house, he perverted my view of God's universe, he whispered accusations against this person or that person, this group or that group; he claimed they were followers of the Antichrist. But when he left my house to go out into the world and spread devastation, when I saw him rising like a shadow over all the earth, not a shred of hatred was left in my heart. He had nothing more to teach me, except his own emptiness, his shadow-nature. By revealing himself as pure shadow he bore witness to the Light, the great penetrating, searching, unveiling, unmanifesting, and healing light of God now breaking over the world. The breaking of that Light is eternal. It is at the core of every moment. The end of the world lies hidden in every moment. The termination of the cycle, the dissolution of all things, the passing away of heaven and earth, the dawning of the new heaven and the new earth, is always there, in time present pregnant with time future, where the whole creation groans to be delivered—until *now*. 'When a man rejects error and embraces truth,' said William Blake, 'a final Judgement passes upon that man.'

The proper use, the specific spiritual practice of apocalyptic times is: To let everything be taken away from us, except the Truth. When Blake cried, 'Whatever can be destroyed must be

destroyed!', this is what he meant. Whoever can—with the aid of
Heaven—not reverse, but simply *resist* the tremendous centrifu-
gal, scattering, attenuating and sinking forces active at the end of
the Aeon, will find that all the dross in his soul, all the sin, all the
spiritual heaviness and intellectual darkness of the latter days, has
been stolen from him by the Antichrist. He is welcome to it.
[Antichrist] will burn out sorrow with sorrow and fear with fear,
since only in the presence of God's Mercy can we face the full
depth of the sorrow and fear all of us feel at the end of the cycle,
and witness their essential emptiness. If we can resist despair in all
its forms, including violent panic, cold-heartedness, and false
luciferian hope, then, after all the karmic residues of the entire
cycle have been torn away from us, there we will stand, naked, in
utter simplicity, before the face of God. This is the meaning of 'for
the sake of the elect those days shall be shortened,' and 'the meek
shall inherit the earth.' Whatever in us 'crystallizes', to use one of
Schuon's favorite terms, in the presence of Absolute Truth, will
be 'gathered into the barns' where the fertile potentials, the 'seed
corn' for the next Aeon, are stored. 'He that shall endure to the
end, the same shall be saved': he shall be *saved up*....whatever in us
resists the temptation to flee God by taking refuge in chaotic
dissolution—to hide from the destruction of matter, or the fear of
this destruction, in matter itself, which is one meaning of 'they
shall pray for the mountains to fall and cover them . . . will enter
the feast of the Pirs, the Shaikhs, the Tzaddiks. . . .

When you find yourself in a state of fear or grief over the evil
of the world, the degeneration of humanity and the ruin of the
earth, know that this evil, ruin and degeneration are nothing but
the mass resistance of the world to the impending advent of the
Mahdi, the Tenth Avatar, the Messiah—and that the fear or grief
you are presently experiencing are *your way of participating in that
resistance.* Knowing this, simply stop resisting Him, and let the
Messiah come. Stop trying to maintain the world in existence by
the power of your ego; let it go. Let it end. Let your ego end.
You've been fighting off the Messiah: cease hostilities now, 'resist
not evil' (which is how your ego experiences Him), lay down
your weapons, and let Him break through 'the clouds of heaven',
the clouds of individual and collective egotism which have separated earth from its divine Source ever since the fall of man.

On Mummification, Incorruption, and the Resurrection of the Body

I've always suspected that the Egyptian practice of mummification was an attempt to extend, into a later age and by artificial means, the grace of an earlier and holier time when incorruption of corpses was more common.

Of course you die when your body wears out, nor does the resurrection of the body mean the reanimation of your corpse, some of the atoms of which may now be parts of the bodies of mites and mules and other human beings. Incorrupt or partially incorrupt corpses are signs in this world of a higher incorruption—of the kind of eternal union of body and soul which is not possible in this world made corrupt by sin. The Shi'ite theosophers define a 'conscious being' in terms of the polarity between spirit and body, or invisible Source and visible manifestation. Every spiritual reality has a material manifestation, just as every material object hides a spiritual essence; this is called 'the Science of the Balance'. When a person dies, they don't just float around as a soul without a body; rather, the dissolution of the body translates the soul to a higher 'octave', where the polarization-and-reunion of soul and body recurs on a subtler level—and 'subtler' here means 'less abstract; more concrete.' This translation is not always immediate, however. Any passions and impurities which beset the soul at the time of death intervene between the full reunion of soul and body—of the invisible essence of the being with its substantial manifestation. Postmortem purgation is the struggle to be fully 're-born', to attain Incarnation on a subtler level—that of the 'Earthly Paradise'. This is, precisely, the Resurrection of the Body. And to a limited degree, the purification of the soul in this life may in some cases produce a more perfect union of soul and body in this world, and so result in physical healing. This is how Jesus Christ healed the sick.

On Polytheism as Monotheism's Decadence

I believe that there is always a degeneration, during the course of a given cycle (which various divine revelations are sent to partially reverse) from the sense of the Names of the One God as direct expressions of His Reality, or of our necessarily limited attempts to understand that Reality, to the level where these Names are perceived as semi-independent 'archangelic' hypostases (like the Zoroastrian *amesha-spentas*), and from there to the level of apparently independent gods, which is polytheism. (The archaic Orphic tradition saw the gods as representing divine hypostases and ontological levels; the Classical Greeks, apart perhaps from a few educated in the Mystery Schools, tended to see them as independent individuals, capable of jealousy, collusion, conflict, etc.) And in our time, perhaps the sense of God's attributes has fallen all the way to the level of data: facts are innumerable, and only facts are real.

On Prayer as Pure Act

The Aristotelian concept of Pure Act is attributable to Necessary Being alone, that is, to God. In God all possibilities are eternally actualized; the best His creatures can do is move from potency (possibility, potentiality) toward act (actualization), and the closest we come to Pure Act is indeed prayer. We can see this even more clearly in Shaivite Hindu doctrine: Shiva, the Absolute Witness, as Pure Act, is immobile; it is his Shakti who is both dynamic (engaged in what we would think of as action, the motion from potency to act) and passive with regards to the Pure Act that is Shiva. To become enmeshed in action beyond what God wills for us is to lose self-determination, to become a plaything of contingent forces. In some ways the powerful men of this world have less freedom than any of us; some day this will be made clear to them, incontrovertibly.

The Eye of the Buddha

The Mahayana Buddhists say: 'The voidness of things (*shunyata*), their lack of any intrinsic self-nature, is inseparable from the suchness of things (*tathata*), their quality of being precisely as they are in this moment.' They also say: 'The realization of emptiness (*shunyata*) is the presence of compassion (*karuna*).'

Shunyata voids my definitions of things. Before *shunyata* is realized, I sure as hell know what *I* am, and I know what *those* things are, too: *that* thing is a tree; *that* thing is a rock; and *that* thing, over there, is a son-of-a-bitch.

But after *shunyata* is realized, after my definitions and evaluations of things are voided, all things stand out in their intrinsic 'suchness', their *tathata*. Each thing is revealed as bound to the limitations of its form and fated to suffer the destiny of that form. Thus *shunyata*, in revealing the suchness of things, also reveals the suffering of things, while at the same time negating my evaluations of things. In accomplishing these two *paramitas*, it subsists as compassion—not *my* compassion, but *intrinsic* compassion.

This is the Eye of the Buddha.

On Christianity
and Eating Disorders

Food, sex, and a certain degree of aggression all make human life possible, but they can also distort and ultimately destroy it. From a Christian perspective, the Eucharist, *panis angelicus* ('angelic bread') is, as Jungian psychologist Marion Woodman pointed out, intimately related to the psychology of food and eating, and also the distortions of it: not in its essence, which absolutely transcends psychology, but in terms of the psychic reverberations of the Eucharist within the collective mind of a Christian or even a post-Christian society. God is the primal 'food' of all living things (as in 'take and eat: this is My Body'), while, from another point of view, He is also the ultimate Eater. (This is also a major theme in the Upanishads, one that has everything to do with the meaning and efficacy of sacrifice, given that all true sacrifice is ultimately self-sacrifice.)

Eating disorders are among the central signs of that René Guénon called 'the reign of quantity'—the tendency, in these latter days, for everything to be defined only in quantitative terms, such that the inherent *qualities* of things drop out of sight. In the words of William Blake, 'More! More! is the cry of a mistaken soul; less than All can never satisfy man'—and who, like Oliver Twist, has ever asked for a 'second helping' of the Holy Eucharist? In that Bread, the All, the fullness of the Godhead, is in the smallest portion; the Eucharist is thus the ultimate seal of the reign of pure quality, of the return of humanity to the Garden of Eden.

Why do so many anorexic teenage girls who have starved themselves to walking skeletons still see themselves as fat when they look in the mirror? They are fat with the self-involvement, the narcissistic self-worship imposed on them by profane society. (According to one recent study, 40% of American teen-agers now want plastic surgery!) And what is it that bulimics can't stomach? The mores of consumer ('eater') society—the reign of quantity itself.

Grace is given freely. God ploughs and harrows the field, plants the seed, cultivates the wheat, harvests it, threshes it, winnows it, grinds it in those mills of His that grind slow but exceeding fine, kneads the dough, adds the leaven, bakes it, and serves it up (these being steps in a strictly alchemical process). He may even go so far as to put it in our mouths for us. But we still have to chew, swallow and assimilate it. (We also *are* the grain, the bread, by virtue of *communion* in all its meanings; the psycho-physical element in us, the lower self symbolized in the Old Testament by Leviathan the sea-beast, is roasted and served up on the Father's table at the Wedding Feast of the Lamb.) There is a Russian fairy tale that goes like this: Once there were three brothers who were the laziest men in the world. They refused to work for their daily bread, and so (not surprisingly) they ended up pretty hungry. One day they came upon an apple tree, but could not imagine climbing the tree to pick the apples, or even shaking the branches. Then one brother had an idea: 'If we lie on our back under the tree with our mouths open, surely apples will eventually fall into our mouths, won't they?' So that's what they did.

After lying there for several hours, however, a second brother began to have certain misgivings: 'But even after the apples fall into our mouths, we'll still have to *chew* them, won't we?' And with that thought the three brothers abandoned their over-ambitious plans, got up, and walked away from the apple tree. Which is to say: The full and balanced assimilation of freely-given Grace takes real spiritual work, this being another varia-tion on the theme of faith and works, or of predestination and free will. *Receptivity requires sacrifice* (cf. the Magnificat). And this is the built-in contradiction of the sin of gluttony: to 'wolf' your food like the wolf wolfed Little Red Riding Hood's grandma is to make receptivity impossible: the more you over-eat, the less real nourishment you receive.

Just as God, or Primordial Man, sacrifices Himself to create the universe (a motif found in many myths, the Norse myth of Ymir for example)—He is literally butchered, dismembered to produce that collection parts, or partial views of reality, which is this fallen world—so Christ is butchered to redeem it; by eating His flesh and drinking His blood we become the sons and daughters, the 'flesh and blood', of our Father. By that dismemberment we are remembered—remembered by God, and in Him.

The sin of gluttony (*gula*, related traditionally to the expansive, kingly, generous planet Jupiter) is nothing less than an attempt to *incorporate the world*. The king's role is to give law and order and largesse to the people, drawing on the storehouses of the Eternal King, thus making of them a kingdom. He is the one who owns the larder, who guards the storehouses, who keeps the bread; he is Our *Lord* (an Anglo-Saxon word that means 'loaf-ward'). But if he fails in this, he will eat the people instead; he will eat his children like Saturn did. ('When Jupiter and Saturn meet/O what a crop of mummy wheat' said Yeats—there's a riddle for you). When the vertical dimension, the reign of quality, is lost, this *world* becomes our all in all. But we are alienated from that world, cut off from nature, unable to squeeze the full juice of life out of that maddening, smothering, all-encompassing stone. God is the priest who officiates at the marriage of man and nature, man and his *shakti* or (in Blake's words) his *emanation*; without His blessing, our union with nature is mere fornication; our world, our very life, proves fickle, unfaithful. And we know that the Great Mother Nature (Little Red Riding Hood's grandma) will eat us up in the end, when we are sealed at last into that great sarcophagus ('flesh eater'). So our only recourse seems to be to eat her instead; the only way we can imagine overcoming the shadow of alienation that has fallen between us and our lives, us and our world, is to devour that world. (Obviously this has everything to do with our present environmental destruction.) But, of course, those who devour their world, and their own life, have no life or world left. Just as the drinker is always dry, so the glutton is always starving: starving for the Bread of Life.

In some TV 'reality' programs (what a misnomer!), people attempting to overcome certain vices will confess to total strangers on the street. The goal is for them to confess, if possible, to the whole world. But the world cannot forgive us, especially since our primal sin is the sin of dissipation, which is what this world actually *is*. Confessing to the millions simply tears us into a million pieces. Confessing to God, however, *recollects* us; His broken bread unites us; His wine makes us sober. At the Wedding Feast of the Lamb at the end of time (the end of craving, the end of desire), it is Leviathan the sea-beast that is slaughtered, cooked and served up to the guests attending the eschatological banquet; Leviathan is the regime of nature, the passions, the *nafs*. Instead of the passions eating us, it is now we who are invited to eat them; the final reward of faith and good works is to hunt, kill, slaughter, cook, season and feast upon the sub-human energies of nature, and in so doing, by spiritual assimilation, transmute them into the substance of the Eternal Human Form.

On Hiding
or Admitting One's Sins

ASCETICISM is necessary but not sufficient for the spiritual Path. We need to go to war with the passions, with what the Sufis call the *nafs al-ammara;* but if we fight this battle with our own resources and in our own name, then we will lose. The ego with its self-will cannot overcome the passions, because self-will is a passion in itself, perhaps the greatest one. Only God can conquer the *nafs*—which is why, if we wait until we have subdued the lower soul before practicing the presence of God, we may well wait forever. We want to appear before God bathed and manicured, in our suit and tie; and this is a worthy desire in the face of God's majesty. But the other side of this desire is that we foolishly believe we can hide our impurities from Omniscience; we believe this because we want at all costs to avoid the *shame* of the encounter between our sinful souls and *al-Quddus*, the Holy. Yet that shame is in itself a spiritual virtue. If the Physician is to heal us, we will have to 'come clean' with regard to our disease.

True and False War;
True and False Peace

There is a true peace and a false peace, just as there is a false war and a true war.

The false peace is the peace of those who follow the changing whims of the world, and do whatever it commands, unconsciously, without question.

The false war is the war of the ego against other egos, and against itself. It is the result of inner division, illusion, selfishness, hypocrisy, and it inevitably creates war in the outer world.

Those who have fallen into the false peace of obedience to the world will ultimately be led into the false war, the conflict of ego against ego, and of the ego against itself; in that false peace, there is no peace.

The true war is what the Prophet Muhammad called 'the war against the self'—not the war of the ego against itself, which is the hopeless struggle of what is inherently divided to assert its own self-willed unity, but the war of the Spirit of God within the Heart against the ego that wants to deny It; hide It; appropriate It; destroy It.

And the true peace is the peace of those who have reached unity of character, based on trust in God and certainty in the presence of Truth—the peace of the victory of the Spirit over the ego, the lower self.

But those who have reached inner peace and security, who no longer move and flow with every current of the world, will be attacked by the world—not because they have abandoned their true selves to go to war with it, but because they have found their true selves, and so are no longer obedient to it. Because they have arrived at peace, and certainly, and security, and have come to rest, the shifting currents of the world will beat against them.

So is there no true peace after all? If there is still a 'world' beating against a 'true self', even the self-at-peace, then there is still the shadow of an ego. Only when there is no world, no ego, but God Alone, will the beating of those waves against the self which has attained inner peace finally come to an end. At that point those waves are no longer the waves of the world, but the waves of God—and the self, which now holds for itself no identity other than God, no longer stands against those waves, but rides them, in what manner, and to what destination, God wills.

PART FIVE

The Last Things

Personal
Immortality
vs. Liberation

Roughly speaking, the western, Abrahamic spiritualities—Judaism, Christianity, and Islam—tend to emphasize personal immortality in their accounts of the ultimate destiny of the soul, while eastern religions, notably Hinduism and Buddhism, see the persistence of individual identity as something to be liberated from, not something to seek as if it were the highest goal of the spiritual life. Even the most exalted paradises are still part of the realm of name and form, and thus subject to mutability and decay. When the soul exhausts its favorable *karma*, or when the entire formal universe itself is reabsorbed into the Formless Absolute, the personal immortality of even those souls who have attained posthumous bliss is ended. To say it another way, Liberation is virtual within the western religions, while immortality is explicit; within the eastern religions, the reverse is the case.

Are the notions of personal immortality and ultimate Liberation fundamentally opposed? I don't think so.

As time-bound, contingent beings, we are mortal; as the Buddhists never tire of pointing out, no relative and compounded state, either incarnate or posthumous, can be eternal. In the words of the Qur'an, *All is perishing except His face.* Yet God *is* eternal, and as such He sees all things *sub specie aeternitatis.* What to us is a temporal event is to Him an eternal form. In time we pass and die, and are reborn in higher or lower worlds, only to pass and die again. But if ever we once were, in God we eternally are; this, not some indefinite persistence through time of our relative, compounded selfhoods, is personal immortality. Yet this immortality is not a property of ourselves as self-reflexive subjects, but of God, or rather of our eternal archetypes in God. We are immortal, in other words, not in terms of our experience of ourselves as continuously existing beings, even in blissful communion with God, but in terms of *God's experience of us* as reflections or aspects of Himself.

Hinduism posits three planes of being: *kama*, *rupa* and *arupa*. *Kama*, as 'desire', is the principle of time, which in terms of the consciousness of sentient beings is based on the desire to possess or avoid this or that experience. *Rupa* is 'form', and it transcends time; that which is perfectly itself doesn't need to seek or avoid something in order to complete itself or protect itself. And *arupa*, which transcends *rupa*, is the Formless. God is Formless, yet within Him is form, to which He lends, not His self-sufficiency, but His immortality, or a certain degree of it. The world of timeless forms or Platonic archetypes is not self-subsisting; it depends for its existence upon the Formless Absolute. Nonetheless, it still transcends duration as we experience it. Even though it does not partake of the perfect eternity of God, it remains 'relatively eternal' in relation to passing time.

As a reflection of God's eternity, *rupa* is equivalent to the Greek concept of *aion*. 'Aeonian time' is a portion of passing time considered as a single eternal form, much as a year can be represented visually on a single page as a cycle of twelve months. To say 'time is the moving image of eternity' (cf. Plato's *Timaeus*) is to posit aeonian time; this is the level of personal immortality. What in passing time is a lifetime of experiences and choices, in aeonian time is a single eternal form, enthroning a single, complex, sovereign choice, a timeless choice like that of the angels, who chose to obey or disobey God 'before time began'. But because of the nature of eternity, which underlies the moments of our lives rather than literally coming before or after them, this eternal choice does not strictly predetermine our daily choices in passing time, since it can with equal validity be considered as the final sum of them.

Just as time (*kama*) is a moving image of the eternal forms (*rupa*), so these forms are dependent upon the Formless (*arupa*), without which they would be caught in the stream of passing time, and therefore not be eternal. In Buddhist terms, the fact that the Formless is the source of all forms is the principle of the 'voidness' of forms, their absence of self-nature—and that which has no self-nature can neither persist nor pass away.

We are immortal in our limited, human forms because, due to God's vision of all forms *sub specie aeternitatis*, whatever has once existed necessarily exists eternally; we are not simply subsumed or dissolved in the Divine Nature. Yet this immortality is real only by virtue of our total Liberation from self-reference and the illusion of self-existence, the factors upon which time and mortality are based—a Liberation which seems, from the temporal point-of-view, like a goal to be attained, but which from the standpoint of God's eternity is simply the real nature of things. As the Cha'an Buddhists say, 'from the beginning, not a thing is.'

If this eternally-existing Liberation is not realized, we seem to wander from form to form in passing time, lost in the illusion of *sangsara*, always dying yet never cleanly annihilated, always becoming yet never quite able to be. This *sangsara* is based on our craving to *be our own experience of ourselves*—a craving which can never be satiated, since an experience can never be objective to itself, and thus can never be fully experienced; it remains a mere virtual reality. Our immortality as limited, contingent forms cannot be based on our own experience of ourselves, but only on God's experience of us as eternal reflections or aspects of Himself, according to which—from one point of view—He is the only Experiencer, yet equally no experiencer at all, since in essence there is nothing other than That One for Him to experience. The eternal reality of a human form is only realized when that form dies to itself, when the ego is annihilated and the Eternal Divine Witness, the *atman*, is unveiled; this is one meaning of the *hadith* of Muhammad (peace and blessings upon him), 'he who knows himself knows his Lord.' This is *moksha*, Liberation. Personal immortality, in other words, exists only by virtue of the total objectification of the human subject before God, the annihilation of its illusory self-nature, which is complete liberation from ego. Where self-identification is ended, the human self remains as one among God's infinite possible Self-experiences—He who, in another sense, is absolutely beyond experience, since He is 'One without a second'. This is the *fana* and *baqa*, the 'annihilation and subsistence' of the Sufis. It is the import of those Buddhist *tankas* (roughly equivalent to icons) where this or that enlightened sage is represented in terms of his or her eternal form, empty of self-nature yet available to the non-yet-liberated as an intercessor, a channel for the transmission—virtually at least—of Perfect Total Enlightenment. It is what St. Paul meant when he said, 'it is not I who live, but Christ lives in me,' and the real import of the words of Jesus, 'he who seeks to keep his life shall lose it'—i.e., the attempt to exist as one's experience of oneself leads to endless wandering in *sangsara*—'but he who loses his life, for My sake'—for the sake of the *atman*, the indwelling Divine Witness—'shall find it'.

There is no essential contradiction, then, between self-annihilation, personal immortality, and ultimate Liberation from the round of becoming, these being three different aspects of the same Reality. Approaches differ; the Truth is One.

The Solipsistic
Denial of the Afterlife

The experience of complete unconsciousness is by definition impossible. No human being, no sentient being, has ever had such an experience. To put it in ontological language (given that, in the Absolute, Being and Consciousness are equivalent), *absolute non-existence cannot exist.*

Those who believe that death ends everything in effect believe that their own annihilation entails the annihilation of all consciousness, the end of the universe. This belief is the inescapable obverse of the tacit belief that 'only I exist'. If I am the only conscious, existing being, if all other objects and beings are simply my own dreams, then of course my annihilation entails the annihilation of the universe; of course death ends everything.

But unless I am a complete solipsist, I will have to admit that other conscious beings exist. And if they do, then how could the end of *my* consciousness annihilate *their* consciousness? 'Of course other conscious beings will still exist after I'm gone,' says the denier of the afterlife; 'I simply won't be there to be conscious of them.' To say this, however, is to claim that one can, as it were, 'inhabit non-existence' in such a way as to negate all being. It is based on the attempt to imagine oneself as not existing, which can't really be done. One imagines a darkness, a blankness, an emptiness which can never be complete, otherwise one would be unable to distinguish it from other imaginings, and would therefore be unable to imagine it. One envisions one's death as the everlasting 'survival' of one's personal oblivion, a fixed point of private destruction which leaves no room for other conscious beings and their experiences. But if I have truly ended, then my personal oblivion has also ended; I am no longer there to negate the consciousness of others. And if

my own absolute oblivion cannot be established by my death—
and if the consciousness of others cannot be negated by it—
then death is not the end.

Someone who is incapable of letting go of his own self-concept, because he totally identifies with it, is forced to believe
that outside this self-concept nothing exists. What he is actually
experiencing in his dreams of non-existence, however, is the
relative vacuity of that concept itself, the relative non-existence
of the ego, which is a mere shadow of reality. (The ego always
tries to become immortal by taking refuge in an imaginary and
impossible realm where it can, as it were, feed off its own non-
existence, where true Being can never dawn upon it to expose
that non-existence and annihilate it.) Death, however, entails
the end of one's self-concept, the end of that vacuity—not the
final end, otherwise physical death would be equivalent to final
Liberation, but certainly the abrupt removal of many of the
materials, both physical and psychic, that one has used to construct one's self-image. What death ends, then—or what death
challenges—is the obscurity and relative non-existence of the
narcissistic ego. It is not the annihilation of Being, but the beginning of the end of all could obscure the face of Being. What dies
in a good death, and more particularly in the spiritual death that
is self-transcendence, is the fantasy that non-existence is somehow possible. (I say 'a good death' to distinguish it from an *infernal* death, which is a willful plunging into the fantasy of non-
existence in order to escape the radiant face of Being which has
suddenly dawned upon someone who has spent his or her life
denying it.) Death is not the annihilation of Being, but the annihilation of the illusion of that such annihilation is in any way
possible. In the words of priest and poet John Donne: 'Death,
thou shalt die.'

What is Death?

Theosophic Meditations
on the Passing of a Friend

Our friend Scott—sometime secretary to Prof. Huston Smith, Platonic/mathematical philosopher (though he would probably deny it), Shadhili *fakir*, follower of Frithjof Schuon and the 'Traditionalists', and convert, during the last year of his life, to Eastern Orthodox Christianity—passed away in November of 1998. In him we lost a true spiritual companion.

He was actually baptized in the hospital; and though I am a Muslim and a dervish of the Nimatullahi Order, I was allowed to be 'acolyte to the acolytes' during both his baptism and his reception of the sacrament of the sick. I spent many days visiting him in the hospital, and was present in his room up to an hour and a half before his death. (My wife Jenny was there the whole time.) As soon as he died, my wife, myself and Scott's priest simultaneously recognized that we now had an advocate in the other world.

As Scott was dying, I realized that it is possible for someone who is suffering profoundly to burn up other people's karmic impurities in the fires of that suffering. I believe that Scott did this for me, and for several others. This is essentially Charles Williams' notion of vicarious suffering, and I am told that Eastern Orthodoxy has a similar doctrine. What a spiritual Master can do during his life, some spiritually advanced souls can apparently do at the point of death.

I saw that death, like birth, comes in waves: contractions. I understood this because I in some sense 'midwived' Scott's death; at least that's how I experienced it. I felt the contractions. It was as if, as I meditated beside his death bed, I was allowing him to die 'through' me, as I spontaneously visualized his soul ascending to Paradise.

My sense of Scott's death on the subtle or animic level was distinct from my experience of the journey and destiny of his immortal soul. During one time when he almost died—he temporarily stopped breathing—I felt the 'etheric field' in the room around him wobble. And the aura of his body after death—in no way identifiable with his soul, but more on the order of a psycho-physical residue, identified by René Guénon with the Hebrew *ob* and the Roman *lares*—was like a sort of crystalline mist.

During the Eastern Orthodox sacrament of the sick, administered the Wednesday before he died, the last day he was really conscious (he passed on at 1:30 next Saturday morning), everyone around his bed felt the heavens open. The priest, Father Stephan, was as amazed as the rest of us; he said had never experienced anything like it. Heaven came down for Scott, then lifted again. After that point his relationship to his physical form was extremely tenuous. His body was still breathing, but his spirit had already largely departed.

After the exaltation of the time immediately before and after his death, the mammoth job of cleaning out his incredibly cluttered apartment and distributing his vast library of spiritual books began. A number of Scott's friends, old and new, and several people from his church (St. Nicholas Russian Orthodox Church in San Anselmo, California) including Father Stephan, made up the work party. I took responsibility for the rescue and distribution of the books, which may have numbered as many as 7,000, including those in storage, many if not most of them spiritual classics. The apartment was in a terrible mess, since for several years he had been too depressed and physically ill to take care of it; rats had even begun to eat the books. Yet everything in it was a jewel, a treasure. This is why I say that the act of cleaning out his apartment was like Christ harrowing Hell. The books concretely symbolized the souls of the righteous in Hades under the Old Dispensation; Scott's death, like the crucifixion and resurrection, inaugurated the New

Dispensation. Consequently the physical work we did was also, by virtue of its inescapable symbolism, a spiritual act.

I learned through this work that the distribution of the personal effects of the deceased both symbolizes and concretely aids in the distribution of that person's psychic legacy—the accidents of his or her personality, including the imprints of lived experience, as well as influences from the dead, the living, and the ancestors: everything which that person has collected during life to weave the psychic garment of his or her immortal soul. It is the distribution of this legacy which constitutes metempsychosis; the soul does not literally reincarnate, yet both the living and the unborn can inherit elements of the personality, memory, worldly destiny and possibly even spiritual merit (or at least its psychic reflection) of the deceased.

The distribution and purification by the living of these personality-elements of the deceased through a spiritually-based grieving process, aids in the journey of the soul after death. I must emphasize that it has nothing whatever to do with the *salvation* of the soul, merely with the ease or rapidity with which that soul rises to its eternal destiny; at least this is true in the case of a sanctified soul.

It is this level of things that's represented by the Orthodox doctrine of the 'tollhouses', so reminiscent of the Egyptian Book of the Dead, as well as the of the doctrine of the *bardo* of the Tibetan Book of the Dead, particularly the Second Bardo. The tollhouses, mythically or symbolically conceived of as something on the order of customs sheds, are stages in the soul's progress in after death, 'borders' which cannot be passed until certain pieces of earthly 'baggage' are left behind. They are thus the rough equivalent of the Roman Catholic Purgatory. They do not, however, possess the force of dogma in Eastern Orthodoxy, and many Orthodox theologians deny their equivalence to the Catholic Purgatory, which they consider to be heresy; others would dispense with the tollhouses entirely. And, certainly, to believe that the passage of the soul through the tollhouses has anything to do with its salvation—as if a 'failure' at one of the tollhouses could actually result in its being damned—would be

heresy indeed. Their reality is on a lower level than that of salvation or damnation; they are psychic rather than spiritual. Nonetheless, it is important for the living to help the souls of the
departed negotiate them, basically through loving detachment,
through letting go.

The three classical stages of the mystical path are Purgation,
Illumination and Union. The tollhouses relate to Purgation; the
spiritual traveler who 'dies before he is made to die' (in the
words of the Prophet Muhammad, peace and blessings be upon
him) passes through the tollhouses in this life. Purgation does
not *cause* Illumination, or Illumination, Union; to believe this
would be to believe that the soul can attain union with God
entirely through its own efforts. Rather, it would be better to say
that virtual Union manifests as Illumination, and virtual Illumination as Purgation, which is the same thing as saying that the
perspective of God, or Grace, supersedes and embraces the perspective of man, or works.

According to the *Bardo Thödol*, the *Tibetan Book of the Dead*,
which is based on the paradigm of reincarnation, there are three
bardos or time-periods in the after-death state (the Fourth Bardo
being earthly life). The First Bardo is the moment of death, when
all form is transcended and the light of the Void or the Original
Mind shines clear; the Second Bardo is the period when the psychic material released from the dissolving subjectivity of the
deceased dawns in a series of archetypal forms; the Third Bardo
is the period when the soul, if it has failed to achieve immediate
liberation in the First Bardo, or proximate liberation (probably
to be identified with Paradise) in the Second Bardo, nor yet
fallen into one of the Hells, chooses its next homeland and its
next parents in the process of seeking rebirth. To me, as a Muslim who does not believe in literal reincarnation, this can simply
mean that those the Qur'an calls the 'foremost' attain direct
annihilation in, and subsistence through, God after their deaths;
those designated as 'on the right' attain Paradise; and those 'on
the left' fall into the Fire. As for re-birth, it might be admissible
to say that those who identify almost completely with their own
subjectivity, though they have committed no major sins, may

seem to 'follow' that subjectivity, that 'psychic legacy', after their deaths, as it is inherited by the already-living, the newly-born, or those about to be born. To the living souls who have appropriated elements of the earthly experience of the deceased, it may seem as if they are actually the dead one reincarnated, while in reality the deceased soul is in a kind of ghostly 'Limbo'—which Dante, we should remember, situates in the 'highest' region of Hell—unable to let go of the accidents of its former personality, and therefore attempting to re-experience earthly life by means of them, as they come to be attached to a new, and unique, human incarnation. (It should go without saying that those among the living who become objects of the attention of ghostly souls in Limbo are themselves likely to be over-attached to their contingent, temporal personalities—wanderers in the wilderness of *sangsara*.)

Scott's death and funeral were the First Bardo; the cleaning of his apartment and distribution of his books were the Second Bardo; and the re-consolidation of our web of relationships, after his loss, under changed circumstances, was the Third Bardo—which has included the arrival of a new acquaintance, who seemed to have come directly through the 'hole' left by Scott's passing. Mysteriously, this person—who turned out to be a failed spiritual struggler eaten up by demonic influences—appears to have incarnated those very personality-elements that Scott repressed during his life, as if he had been born directly from Scott's unconscious—as if, in a metaphorical sense, he were Scott's reincarnation. But in actuality he was a dark soul attracted to the spiritual light emanating from Scott's death—a psychically traumatized and devastated individual with a background in witchcraft, Gurdjieff and Native American shamanism, vulnerable to every kind of 'wandering influence'. To the degree that Scott had unfinished business, powerful and deeply-buried wishes that he had been unable to live out because he could not bring them to an effective center, this man acted as an unwitting host to these wishes. (The grief process, by the way, seems to straddle the Second Bardo and Third Bardos: the Second Bardo corresponding to the pain experienced by his living friends in the process of letting him go, compensated for—in Scott's case—by our reception of the radiance of his own act of letting go, his faithful and willing sacrifice of his earthly form—and the Third Bardo corresponding to that different pain, harder in many ways, the pain of receiving new earthly life after Scott's passing, a life which we could not fully accept without a deepening realization that, in terms his earthly life, he was really no longer with us.)

So it is certainly true that whenever something or someone dies, something or someone else is born, whether or not we have the insight, stamina and detachment to consciously accept it. But it is just as clearly true that this new something or someone is not a literal reincarnation of the person or thing that has passed, but rather a new and unique advent, a fresh act of Divine Providence. In Scott's case, the man who came into our lives after his death was the 'scapegoat' who carried away Scott's unlived life, or a large portion of it, so that the new era could dawn for those who loved him—though for some who were unable to let go of him (one person resorted to mediums after his death, who told her that Scott was not in some other world with God, but invisibly present in a quasi-material way), no new era arrived until much later.

The First Bardo is analogous to the Hindu *Deva-yana*, the Second and Third Bardos to the *Pitri-yana*. The Deva-yana or 'path of the gods' is the afterlife course which leads to final liberation through the Door of the Sun; the Pitri-yana, or 'path of the fathers' (the ancestors) is the course which leads to existence as an ancestral spirit, and ultimately to rebirth. In Neo-Platonic terms, the First Bardo corresponds to *pneuma* or Spirit, the Second Bardo to *psyche*, and the Third Bardo to *soma*, the physical body.

After Scott's death, my attention to him split, between the Way of the Ancestors and the Way of the Gods. I saw that if I gave to Pitri-yana whatever was accidental and contingent in my relationship to and memory of Scott, through a healthy, purgative nostalgia, I lightened his journey, whereas if I deliberately dredged up those memories, or held on to them, or tried to pull them back, through a pathological, morbid nostalgia (equivalent to his friend's attempt to hold on to him through recourse to mediums), then I blocked his way. I saw also that, in terms of Deva-yana, by which I mean Scott's eternal destiny, if I tried to either hide from or to deliberately contact his spiritual essence, I veiled him from me. If, on the other hand, I simply 'stood in wait', without either extending my consciousness toward him or letting myself forget him—a form of *adab* or 'courtesy' that the 'sober' Sufis apply to their relationship with God—then my relationship with Scott was correct. My care to avoid either pulling back his memory on Pitri-yana or reaching for his essence on Deva-yana is part of what has made a true contact with his spiritual essence possible.

Another aspect of the Third Bardo is the sense that he become established at one point as an intercessor, one of the 'people of the graves'. When I visited his grave shortly after his death, I found it easy to communicate with him, because he hadn't quite left. Later it was more difficult; he seemed to be attending to pressing other-worldly business. But then, it became easy again. It's as if he had settled down in his 'assignment' as intercessor—or, on another level, as if he had experienced the 'resurrection of the body'. I could now spontaneously talk with him from time to time, much as I had done in life. (A heavy, necromantic evocation of his soul, however, would have been just as much an act of discourtesy as an uninvited invasion of his home during his earthly life; it would have been more likely to destroy our relationship than deepen it.) And finally, the time came when such communication seemed no longer appropriate—though on at least two occasions I have unexpectedly experienced what I believe to be an intent on his part to communicate with me, both times for a very good reason. On the last occasion, it was as if he made a certain amount of his unlived life available to me in response to a period of intense petitionary prayer, during which I certainly did not have him in mind as an intercessor. This deep invocation of God's will seemed to have attracted certain of his unfulfilled wishes, which appeared to function as *potentia*—latent matter and stored-up power—so as to lend *substance* to a new Form of life coming down from the higher worlds, allowing it to take root in this world. To have deliberately attempted to use the energy of these unfulfilled wishes in order to accomplish this would have been sheer black magic—but what is well forbidden to us, God has both the right and the power to perform. By His Grace, my life-struggle was allowed to aid Scott in his after-death purgation, just as his purgation was tapped to help me in my hour of need. (Whenever I would complain, as a child, about this or that frustration, my Catholic mother would always say, 'offer it up for the good souls in purgatory.' It is only through this experience of an unsought partnership with the departed soul of Scott that I have understood exactly what she meant.)

Both Christianity and Islam teach the resurrection of the body as orthodox doctrine. What does this mean? Are all the scattered atoms of the corpse magically brought together again, even if they have become parts of other living things, including living human bodies? I don't believe that is the case. But neither does the departed exist throughout eternity as a formless ghost.

The Vedanta speaks of a subtle body, the *suksma sarira*, which, according to the *Brahma Sutras*, survives until the final Liberation. Jesus, after his resurrection, appeared in a palpable though 'glorified' body, not a re-animated corpse (though his physical remains were apparently subsumed into it), and both Mulla Sadra and Ibn al-'Arabi, Muslim esoterists, hold that a body is necessary to the soul at every stage of existence.

From one perspective, an individual being can be defined as a polar relationship between its spiritual Source and its formal manifestation, neither of which can exist alone, because they are complementary manifestations of a single Reality. The spiritual pole has precedence over the formal, since Spirit in fact represents this absolute Reality in the mode of polarity with its own manifestation, yet one pole never exists without the other. The dissolution of the material body therefore necessitates a 're-polarization' between Spirit and its manifestation on a different level, thus situating the individual being on a new ontological plane. The material body dissolves, after which the spirit re-manifests a new, integral form for itself on a subtler level. As Scott said shortly before he died, 'What do I have to complain about? I'm simply going to get a new body; that's something I've needed for years and years.'

Who Reincarnates?

The 'popular' view of the doctrine of reincarnation maintains that I am the reincarnated form of someone else who lived in the past, and that I will most likely reincarnate in the future as a third and separate person with a different name, a different form, and possibly a different sex. And yet all three individuals are still claimed to be 'I'. 'I' was Julius Caesar, am now Charles Upton, and at some time in the future will be Another Person. But if 'I' will ultimately be all three of these people, then what justification is there for identifying this 'I' with only one of them, namely Charles Upton, as we do when we say 'Julius Caesar is a past lifetime, and Another Person a future lifetime, of Charles Upton'? Julius Caesar, Charles Upton and Another Person are distinct individuals; Julius Caesar and Another Person are not *really* Charles Upton inside, any more than Charles Upton is secretly either or both of them.

If the reincarnating entity cannot be arbitrarily identified with only one out of its many supposed lifetimes, then we are left with two possibilities. The first is that Julius Caesar simply turned into Charles Upton after he died, just as Charles Upton will eventually turn into Another Person. But if this is the case, then there is no reincarnating entity at all, no inner core of selfhood that first assumed the form of Julius Caesar, later clothed itself in Charles Upton, and will eventually adopt the form of Another Person. All there is is a continuity-in-change with no fixed or essential qualities of its own. And if there is no-one who actually reincarnates, then what becomes of reincarnation?

The second possibility is that there is an entity distinct from Julius Caesar, Charles Upton and Another Person who successively assumes the forms of these individuals, just as you or I would successively don and doff three different suits of clothes. And if this is the case, then Charles Upton is not the reincarnation of Julius Caesar, nor will Charles Upton later reincarnate as Another Person; rather, all three individuals are the reincarnations, or rather the *incarnations*, of a single entity common to all three, yet transcending all three.

The first of these two possibilities represents the Buddhist conception, and the second one the Hindu conception, of the notion of reincarnation. Both the Buddhists and the Hindus consider *sambodhi* (Enlightenment, the Buddhist term), or *moksha* (Liberation, the Hindu term) to spell the end of the process of reincarnation, the final discharge from the prison of becoming.

But who is it that reaches Enlightenment? Who is it that achieves Liberation? In Buddhism, with its doctrine of *anatta* (no-self), there is no continuously-existing entity who passes from lifetime to lifetime, only an ongoing karmic tendency, within the world of *samsara* (illusion), for the false belief to arise that such an entity actually exists. The process of reincarnation depends entirely upon the belief that someone in fact exists who could reincarnate. This erroneous belief, the karmic tendency to this deluded conception of things, does indeed produce a form of experience according to which reincarnation is apparently real. But when we wake from this delusion, when this false idea is dispelled, then not only does reincarnation 'end', but it is clearly seen and fully understood that there never was such a thing in the first place. As the Cha'an Buddhists say, 'all beings are enlightened from the beginning; from the beginning, not a thing is.'

But if an actual experience of reincarnation is possible, then can we not assert that reincarnation has at least a relative reality? Yes and no. The experience of reincarnation is based on an imperfect understanding of who the experiencer is. Reincarnation can never be the experience of oneself *in the actual process of turning into someone else,* since (in the Buddhist conception) the experience of such a change and the experiencer of it (the one witnessing the change, and therefore able to determine that a change has taken place) are one and the same; there is no separate and changeless experiencer apart from the flow of experience. Likewise in the Hindu conception, the Experiencer can never experience Itself as turning into another Experiencer, because the Experiencer is intrinsically *advaita*, 'not two' All that can be experienced is a change in that portion of Its experience which the Experiencer (under the power of Maya) presently identifies as 'myself'. And if reincarnation cannot be the actual experience of one individual self (*jiva*) turning into another, then it can only be the experience of an individual self *remembering* that it has (apparently) been someone else in a past lifetime. Such remembering, many convincing instances of which have been recorded, is adequately explained by *metempsychosis*, the doctrine that psychic qualities discarded by souls at the moment of death—including even memories—can be inherited by newly-incarnating souls, both at the moment of birth and during the process of establishing a stable identity in this world, just as a suit of clothes can be 'handed down' from a deceased relative. As our genetic identities are inherited on the physical level, so our family identity, our community identity, our national identity, our professional identity, our political identity, our religious identity are inherited on the psychic level. Our unique DNA signature is drawn from a 'gene pool'; likewise the soul-material that goes to make up our personal identity is drawn, in a unique and never-to-be-repeated way, from a pre-existing pool of psychic qualities. A stable synthesis of such qualities is necessary to constitute a mature and complete human being; nonetheless, it is our *identification* with these psychic qualities that produces the mass of Ignorance from which we must free ourselves in order to be liberated from the wheel of birth and death.

So the Buddhists deny that there is any actual, reincarnating self persisting over time. And according to the Hindu conception, though it posits the existence of a persistent reincarnating Entity, reincarnation is not for limited, individual beings; it is only Brahman, the Atman, the Absolute Reality, the One Self of All, who reincarnates; He is 'the One and Only Transmigrant'. And yet, as should be obvious, an Absolute and Infinite Reality cannot literally pass from form to form; what is already within all forms, sentient and non-sentient, and yet beyond all forms in the sense that It is incapable of being limited by them, cannot reincarnate. It cannot pass from limitation to limitation. It is beyond all name and form, and absolutely free of them. And so, paradoxically, the realization that Brahman is the One and Only Transmigrant is also the realization that, in reality, transmigration has never occurred, and never will. It is only Maya, the Self-manifesting and Self-veiling power of the Absolute Reality, which makes it appear (to individual beings) that individual beings possessing persistent self-nature, beings capable of reincarnation, actually exist.

So from this perspective we can say, without fear of error, that neither Buddhism nor Hinduism teach the doctrine of reincarnation; rather, they recognize the illusion of reincarnation, an illusion which is ultimately dispelled, to our great good fortune, by Enlightenment, by Liberation. The Buddhist doctrine can therefore be characterized as the *apophatic* approach, and the Hindu doctrine the *cataphatic* approach, to the same basic realization.

But whence this illusion? The Buddhist answer is, 'from the action of interdependent co-creation, based on ignorance'. The Hindu answer is, 'by the mysterious power of Maya, which (paradoxically) is both inherent in Brahman and dispelled by the realization of Brahman.' The Buddhists see 'creation' as the generation of veils of subjectivity limiting and obscuring the Absolute Void. The Hindus see it as an externalization and objectification of the Absolute by the magical, illusion-generating power of Maya. And yet the limitless and undetermined Void which is unveiled when the subjective illusion of individuality is dispelled, and the Absolute Self which is realized when the illusory nature of Its objectifications is realized, are one and the same.

As I see it, the idea or 'experience' of reincarnation is based on the attempt to make sense of things after we have fallen to the error of taking time as an absolute. In one of Zeno's paradoxes, the flight of an arrow when loosed from the bow proved to be logically impossible, because if motion happens in entirely discrete and separate 'quantum' steps, then there is no way for the arrow to pass from step to step, there being absolutely no connection between them, while if motion is purely continuous without being broken into stages, then this very lack of successive stages prohibits any actual *succession* or forward motion. And if motion is logically impossible, then time itself must be an illusion; the Buddhist and Hindu critiques of reincarnation, like Zeno's paradox, are also critiques of 'absolute time'.

And yet, as we know, the arrow *does* fly, or certainly appears to. We all tend to feel ourselves to be the same person from moment to moment; yet we also know that, by any criterion we might apply, we are *not* the same person, that everything in us, everything in our surroundings, our bodies, and our minds, is continuously changing. A given aspect of us may change only minutely, as when we turn our head to look out the window, or drastically, as when we die. But nothing perceivable or definable that in any way relates to our identity is free from continuous change. This inescapable twin perception of authentic identity and inevitable and universal change is the basis of the doctrine of an immortal soul. If we were all change, there would be nothing immortal in us; if we were never subject to change, then we would be no more than fixed, lifeless, *soulless* objects. Buddhist doctrine denies the continuity of the same *phenomenal* self through time, recognizing that our tendency to mistake our phenomenal self for our real Self is the essence of ignorance and the seed of *samsara*, the illusory world perceived and thus 'created' by ignorance. And if the Buddhists do not emphasize the existence of a noumenal Self as opposed to the illusory phenomenal self, it is because, in their practice (or rather their ideal) of avoiding all metaphysical doctrine and speculation that are not directly of use for the purpose of overcoming ignorance and suffering, they realize that to name the noumenal Self is to create a mental image of it as an object among objects, and such a mental object is not the Thing Itself.

In line with its apophatic tendency, Buddhism takes such a mental object to be more of a veil than a revelation. Hinduism, on the other hand, in its more cataphatic manner, posits the reality of the noumenal Self, naming it the Atman, that Absolute

Reality which is the Witness of all the forms of existence while not being one of them, given that 'the Eye cannot (directly and without recourse to a mirror) see itself.' The Buddhists, however, are not limited to the apophatic approach. While always emphasizing that concepts are not realities, that 'the dharma-body is void', they too speak a Great Self or Fair Self (*mahatman* or *kalyanatman*) as opposed to a little self or foul self (*alpatman* or *papatman*), and assert that 'the Self [in the first sense] is Lord of the self [in the second], and its goal'. And certainly such concepts as Nirvana, the Dharma-kaya or the Adi-Buddha point to an Absolute Reality conceived of in more-or-less positive terms.

According to Buddhist scriptures, 'If it were not for the Unborn, the Unmade, the Uncompounded, there would be no liberation from what is born, what is made, what is compounded', and 'There is, monks, a Realm devoid of earth, air, fire and water. It is not the realm of infinite space or the realm of infinite aether or the realm of infinite consciousness; It is the ending of sorrow.' The Unborn, the Realm devoid of form, is not simply the state of a sentient being after he becomes Buddha: it already *is*. (As the practitioners of Zen say, 'If you want to be enlightened, first you have to be enlightened.') When Gautama was on his way from Benares to Uruvela, he ran into a party of young men on an outing. One of them had come with a woman he was not married to, who had stolen all his possessions and disappeared. When the young men asked Gautama if he had seen the woman, he replied: 'What do you say, young men: is it better to track the woman, or track the Self?' They answered that it was better to track the Self, and all embraced Buddhism on the spot. And so the doctrines of Hinduism and Buddhism in this regard, for all their great differences in emphasis, ultimately come down to the same thing.

According to the Buddha, there is no absolute and unchanging self to be identified with the phenomenon known as Charles Upton. According to Shankara, the fact that Charles Upton inevitably feels himself to be himself and no-one else is the natural sign of the Atman, the Absolute Self within him. The Hindu doctrine recognizes Eternity, and understands how Eternity, by

the power of Maya, expressed Itself in time. The Buddhist doctrine utilizes the inescapable and ever-present experience of
time as an *upaya*, an instance of 'skillful means', precisely in
order to *destroy time*: if there is nothing that might pass from
moment to moment, then there is in fact no passage. And when
time is destroyed, then Eternity is unveiled. As the Hindus
express it, 'time, the devourer, ends by devouring itself.'

So the question 'who reincarnates?' ultimately resolves itself
into the question 'who am I?' The answer, in apophatic terms, is:
'I am no-one': *anatta*; in cataphatic terms, it is 'I am the very One':
tat twam asi. But in neither case do 'I', or could 'I', reincarnate.

Reincarnation is the delusion that time, defined as the projection of future experience based on the memory of past experience, represents reality as such. It does not. Who I remember
being an hour or an instant ago is not who I really *was* an hour or
an instant ago. That version of me, filtered through innumerable conditions, sub-atomic, chemical, neurological, psychological, social, and historical, no longer accurately represents the
reality it attempts to transmit; all past memories, due to the
inevitable 'entropy' in the accurate transmission of information,
must inevitably be involved with illusion. Only the *actual* past
moment, not my memory of it, is truly and authentically itself.
And yet, when that moment was not past but present—when it
was an authentic reality, not an imperfect memory—what was
the nature of it? It was, precisely, a present moment filled only
with past memories and future projections based upon those
memories. Even the sense data embraced by it were not perfectly immediate, given that the speed of the nervous system
that converted that moment's incoming sound and light and
touch into my personal experience was not infinite, any more
than were the respective speeds of the physical friction, sound
waves and light waves that carried those impressions to the
doors of my senses. All that really existed—all that really exists
now—is a present moment totally empty of any determined
self-nature, filled with experiences that are essentially *interpretations*, not realities—interpretations, precisely, of the one and
only Present Moment which underlies all phenomenal experi-

ence. That apparent flow of interpretations is none other than Maya, a word derived from the Sanskrit root *ma*, to measure — to measure, and thus to interpret. And that eternal Present Moment, empty of all self-determinations, is the Absolute Witness of that flow of interpretations, just as they are the inevitable expressions of It in space and time. That stream of interpretations, of successive and imperfect versions of the Absolute Reality, is inevitable because the Absolute must radiate infinitely diverse versions of Itself, simply because there is no barrier to prevent It from doing so. And the end of that stream of manifestations and/or interpretations is inevitable too, given that only the Absolute Reality is absolutely real, and what is real must necessarily prevail over what is not. Because the Absolute must radiate infinite versions of Itself, Charles Upton was inevitable; this is the immortality of the soul. Because the Absolute must and will prevail, Charles Upton is nothing: this is *mukhti, sambodhi,* Liberation from the wheel of birth and death.

It may nonetheless seem that, from the Buddhist point of view especially, there could be no such thing as an immortal soul, given that each moment is unique and never to be repeated, and that consequently there is no absolute, individual self that persists over time. It is through directly investigating, by means of the practice of *vipassana* or 'mindfulness', the truth that all things are impermanent in time (*anicca*), and thus that there is no permanent self persisting through time (*anatta*), that Theravadin Buddhism *refutes time*: if there is nothing to pass from moment to moment, there can be no such thing as sequence.[3] But from the standpoint of the undeniable fact that the unique and never-repeated moments of experience, and choice, that go to make up a particular human being have an unbending affinity for each other, that they share certain common qualities that are

---

3. We might be able to get a more concrete sense of the illusory nature of time by attempting to answer the following question: 'Which way is time moving—from the past toward the future, of from the future toward the past?' We seem to move from the past to the future, yet all events, including the very stages of our life, are moving from the future to the past — and what else are we, in temporal terms, but the sequence of events that make up our life?

themselves unique, qualities that no other 'set' of otherwise unique and unrepeatable experiences could ever share in precisely the same way, it must be concluded that human persons are real. We are real not because we have a real, permanent and individual self *hidden inside* us somewhere, but because, by virtue our unique 'suchness' (*tathata*) — which is inseparable from our inherent 'voidness' (*shunyata*), our lack of any persistent, unchanging kernel of self-nature — each of us is uniquely him- or herself, unlike any other. If this were not the case, then Wisdom or *prajña* would not be inseparable from Compassion or *karuna*; sentient beings who are nothing but meaningless illusions, rather than unique manifestations of the Void in the world of form ('form is emptiness, emptiness is form' says the *Heart Sutra*) would not be worth saving. This is the Mahayana view—a view which (in my opinion) is both an inescapable conclusion (in intellectual terms) and an undeniable realization (in contemplative terms) flowing from the Theravadin understanding of *anatta*. This view is iconographically expressed in those Vajra-yana *thankas* that depict individual human beings, like Milarepa or Padma-Sambhava, as eternalized by their own Perfect Total Enlightenment—empty of self-nature, therefore beyond all possibility of passing-away. (It is true that if there is nothing that could pass away there is also nothing that could persist, but eternity has nothing to do with *persistence*: both persistence and passing-away are temporal concepts.) And if time is an illusion—as the Theravada proves, and the Vajra-yana depicts—then unique human persons, defined as unique sets of experiential and intentional qualities, never to be repeated, are eternal: this is precisely where—all appearances apart—the doctrine of the immortality of the soul held by the Abrahamic religions meets the Buddhist doctrine of *anatta*, and is revealed as an inescapable consequence of it: inevitably so, because Truth is One.

Mystery of Iniquity,
Mystery of Mercy

A Meditation on the Occasion
of the Publication of
The Gospel of Judas

When the publication of the Gnostic *Gospel of Judas* was announced, precipitating a flurry of mindless affirmation and almost equally mindless criticism, something began to move in my soul. *The Gospel of Judas* portrays Judas not a Jesus' betrayer, but rather as His confederate in 'setting up' the crucifixion. But this ancient perversion of the Gospel account, which degrades Jesus from the status of the only-begotten Son of God to that of a worldly conspirator and swindler, set up its own counter-motion within me. In the face of the faithless tendency to excuse Christ's betrayer, which is only one instance of a much wider tendency in the post-pedophilia-scandal Catholic Church, and elsewhere, to self-servingly deny original sin, deny the gravity of personal sin, and posit the universal salvation of each individual soul, whether holy or corrupt, I suddenly understood, and became subject to, the Mercy of God—a ruthless, harrowing Mercy which called up all the darkness in my own soul. I saw that repentance, which is instantaneous, initiates an ongoing process of purgation, whereby those faculties of the soul that had been diverted from their true purposes, according to the human form as God created it, are progressively reoriented toward, and remarried to, those purposes.

I also saw that there are certain things in the soul, certain totally perverse tendencies, which cannot be rejoined to their true purposes because they have none. They are purely infernal, purely satanic, and as such have no right to exist. These tendencies cannot be redeemed; they can only be annihilated. As such, they cannot be the objects of the purgation initiated by repentance, if we define this purgation as the redemption of the human faculties. They are, as it were, perpetually in Hell; therefore only the annihilation of Hell itself can do away with them. In light of this revelation, I understood that the doctrine of *apocatastasis*, the restoration of all things in God, which has been so problematical in the Christian tradition, cannot and must not be identified with the salvation of every individual soul on its own plane of existence, which is an affront to the Justice of God, as well as being clearly heretical. The only correct way to understand it is as the reabsorption of both Hell and Paradise into the Absolute Divinity at the end of time, at the point between (speaking in universal, cosmic terms) the passing away of the old heaven and earth, and the coming of the new heaven and the new earth. If heaven and earth can pass away, certainly hell can also pass away—after which nothing remains but the Word by which God names Himself and knows Himself, from all eternity.

We sometimes say that the true greatness God's Mercy can only be understood in relation to the gravity of sin: If we are convicted of sin and understand its enormity, only then we will understand the glory of God's Mercy and Forgiveness. But it is even truer to say that we cannot understand the enormity of sin except in close proximity to Mercy. Sin hides in its own darkness, until the coming of the light. Only then can we see—can we *bring ourselves* to see—the horror of the prison we have just escaped.

God's Mercy is an aspect of God's Truth. Any doctrine which asserts that Mercy absolves us of the duty to confront our own darkness, that God wills to forgive us without our participation, and (as it were) behind our backs, is false. To hope and believe that God will *excuse* us by virtue of an *indiscriminate* Mercy, a Mercy without intelligence and without Truth, is an expression of spiritual despair. To believe this is to believe that if the true gravity of sin were dragged into the light, God could not forgive it. By this belief we attribute to God, under the name of cruel judgment, what is in reality our own faithlessness, our fear that if we were ever confronted with the enormity of our own sin, we—like Judas—would despair of God's Mercy. And this is precisely what Satan wants. Satan is not only the Accuser but also the *Ex*cuser. He excuses our sins in order to suggest thereby that our debts are too great for God to actually forgive, thus denying the efficacy of the Atonement and claiming that the final import of God's Mercy is not to *forgive* our sins but simply to let us weasel out of them through the back door—through denial, through mendacious rationalizations, through self-serving and sentimental false pieties, and finally (when these fail, as they inevitably will) through seeking the 'all-excusing' oblivion of death, as if death were not in fact the *end* of all worldly oblivion, either to our great good fortune or to our irretrievable loss. And as soon as we have accepted the lame and lying excuses Satan suggests to us, blasphemously identifying them with the Mercy of God, then he changes with lightning swiftness from indulgent Excuser to merciless Accuser, tearing down all the cruelly permissive excuses he has just provided us with, and confronting us with the enormity of our sins *in the absence of Mercy*— because to identify God's Mercy with the kind of lame excuse that wouldn't stand up in any court is to throw that Mercy out the window. Moral permissiveness is thus nothing but cruelty; to excuse evil is to embrace despair.

The realization and acceptance of God's Mercy is the very power which enables us to confront the darkness in our souls, and dispel it. Repentance is the acceptance of Mercy. But according to scripture, Judas did not accept God's Mercy; he despaired of it and committed suicide. To imply (as some proponents of *The Gospel of Judas* have done) that Jesus in the Gospel accounts refused to forgive Judas is to cleverly hide the fact that Judas was damned because *he* refused that forgiveness. Jesus commanded us to love our enemies; to say that He had no love for Judas, that He did not *by His very being* extend forgiveness to him, is to accuse Jesus of not practicing what he preached. Of course He extended forgiveness; by His death and resurrection,

He *is* forgiveness. It's simply that Judas refused to accept what was offered. You may freely hand me a $20.00 bill, but if, out of pride (which is another name for despair), I won't take it, then I don't *have* it. It's as simple as that.

In this life, our will is free to choose. In the next life, what we have chosen is fixed; God places His final seal of confirmation on what we have freely chosen, until the end of time. And at the end of the cosmic aeon (not just this earthly one), in the *apocatastasis*, when God takes both Paradise and Hell back into the mystery of His Essence, He also takes back the human forms he has given us, and along with them our human free will. This is the moment when God overwhelms both Heaven and Hell with His Absolute Reality, which is infinitely beyond both of them.

The end of our individual lives is encompassed by the particular judgment; the end of this world, by the general judgment; and the end of universal manifestation, of creation itself, by the *apocatastasis*. To confuse these levels is to fall into grave error. In particular, it is a great and terrible mistake to identify the eternal destiny of the soul on the plane of form with the supraformal eternity of God Himself, realized by the *apocatastasis*. That the doctrine, or the intimation, of such a tremendous and definitive Divine 'event' should be used to deny the gravity of sin, to justify both self-betrayal and the betrayal of God, is an immense blasphemy, and also an immense stupidity—as if a child were to hope that the world would end before his mother discovered that he robbed the cookie jar—as if a serial murderer were to attempt to destroy the whole world in order to drown the tawdry memory of his crimes. God is just; if He were not just, He would not be good; if He were not good, He could not be merciful. And God is not mocked. He does not give us the gift of free will only to let us return it to Him again the moment we realize we have abused it, and that the consequences of such abuse are terrible. If He did, He would in effect be admitting that His creation of the human race with the power of sovereign free will was a joke, a farce, or a mistake. To expect Him to simply and glibly excuse the abuse of His greatest gift is to slander Him, to make Him less than God. (And if rays from the universe-dissolving power of the *apocatastasis* unexpectedly break into Hell, and Heaven too, since they emanate from a degree of eternity that is higher than the dark perpetuity of Hell, higher even than the formal eternity of Paradise, then who are we to deny their power to do so? And who are we to either expect it, claim it, or foolishly rely upon it?)

Hell is a necessary doctrine, a doctrine made necessary by the nature of God and the freedom of the human will. And though we may hope that Hell may be emptied, or that a given soul may be rescued from it, in the mystery of God's Will, to take such action for granted would be like setting one's house on fire and then locking oneself in, in hopes that a windstorm would arrive in a minute to tear off the roof, and rain pour down quench the blaze. It would be the height of presumption, more abysmally and inconceivably foolish than anyone who claims to believe in God has any right to be.

Hell is fundamentally different from after-death purgation. Purgation is for those who have accepted God's Mercy and Forgiveness, who have resisted Satan's temptation (as Miguel de Portugal expresses it) to *choose* Hell because they have despaired of Forgiveness and therefore—in a perversion of Justice, as any Justice cut off from Mercy must necessarily be perverted—think that they *must* go to Hell because they are fundamentally evil. Purgation, on the other hand (as Dante so well portrays it) is a motion of love, joy and gratitude—a love like that for one's Beloved, for whom one would gladly suffer any torment, a love which is also confirmed by the sure hope of reaching the Goal. But if we are not to be overwhelmed by the enormity of our own sins at the moment of death, such that we believe Satan's lie that the only way to wipe them out, or hide from them, is to choose an eternity of hellfire, we had better become familiar with them in this life, and familiar as well with the tremendous Divine Mercy in light of which they are as nothing. If we hope to enter into purgation in the next life, we must begin it in this one.

When Hell is ultimately emptied, it is not emptied by spiritual hope such as those know who are laboring in after-death purgation. Purgation is for the ones who have chosen Mercy, and are in the process of allowing that Mercy, with their willing cooperation, to plough and harrow their souls. Hell is for those who have willed to reject Mercy, and flee from God into the darkness of death, which in reality is only the darkness of their own flight. To fail to distinguish between the saved undergoing purgation, and the damned who are precisely in terror of purgation and in headlong flight from it, is a tremendous error. In Hell, no purgation is possible, neither is it desired. And though Hell may be made necessary by God's justice, in Hell itself there is no justice. The damned have neither the power nor the wish to do justice to one another, nor can they expect any, certainly, from the dark Prince who rules it. Hell is, precisely, a never-ending flight from God.

And this is why, in the ultimate *apocatastasis*, Hell too must die. For a flight-from-God to eternally exist side-by-side with God Himself is a Manichaean affront to His Goodness, His Absoluteness, and His Infinity. And this, precisely, is why revenge is not sweet. For revenge—which in God's terms, is Justice—to be truly sweet, the transgressor must be fully confronted with the horror of his own crimes. And this, precisely, is what cannot happen in Hell, because the enormity of evil can only appear in the light of Mercy, and Hell is closed to Mercy. How often have we seen sexual abusers or serial killers, monsters in human form, interviewed, grilled, cross-examined by various 'researchers' in an attempt to discover *why* they committed such atrocities, *how* they were able or willing to perpetrate such enormous crimes. But these dedicated investigators never seem to get any answers that satisfy them—and they never will, because evil is inherently absurd. Frustratingly, maddeningly, to those trying to make sense of him, the unrepentant Hell-bound criminal can never grasp the evil of the horrible actions he so freely wills, nor can he ever reach this understanding, though we kill him and bring him back to life again a thousand times. Where there should be conscience, there is merely a meaningless, sucking void; where there should be repentance, there is often an insidious, infernal mirth, as if the joke were on anyone stupid enough to look for *meaning* in Hell, and not on the one who deliberately sought meaninglessness, as if it were something to be desired, as if it were somehow a way out. To be strictly accurate there is some truth in both jokes, but the one who sought and found meaning, not the one who fled from it, is the one who laughs last.

But the day will come, when the *apocatastasis* dawns, when all the worlds of form, all the heavenly mansions and the infernal pits, are emptied and dissolved; it is then that Justice will at last be satisfied. On that day, those who have chosen Mercy will be taken, willingly, through and beyond Mercy, while those who have chosen punishment will be dragged unwillingly through punishment, and beyond it. On the day when the gift of free will, eternally sealed in its final choice, infernal or paradisiacal, is definitively withdrawn, because everything other than God is withdrawn, the abysmal flight from God will at last encounter the overwhelming Reality of God, and be annihilated in the face of it. On that day, Hell will no longer be an option. The ultimate consequences of an aeon's flight from Reality will be confronted, and penetrated, and burned away, by the fires of that Reality. On that the day the souls will both suffer the full consequences of their sin—because they are damned—and *understand* the full consequences of it in the light of God's Mercy, *as if* they were saved. The ultimate Mercy and the ultimate Punishment will converge—and that will be the end of them. Only a moment like this, when the exquisite agony of Hellfire is made infinitely more exquisite and intense and soul-shattering by the ruthlessly penetrating Light of Absolute Truth and Mercy, can satisfy Divine Justice. After this point, no flight-from-God, no souls in torment rolling in a lake of fire, will remain. All will be Love; all will be Truth; all will be God. Heaven and Earth will pass away, but His Word will not pass away.

The End
of Knowledge

We are now in a time, late in the Kali-Yuga, when all *collectives*—all self-identified groups, including supposedly traditional and/or esoteric ones—are degenerating. If so, how should we respond? Fr. Rama P. Coomaraswamy transmitted three relevant principles:

> (1) You need to be connected with an orthodox spiritual tradition.

> (2) You need to realize that little is to be expected from this world, which is a vale of tears.

> (3) You need to prepare yourself to be alone.

At the point where the expansion of a spiritual movement has begun to dissipate and invert, the *pneumatics* who have participated in this movement—those who are in it not to make things happen in the material world or establish their identities in the psychic one—are called to *concentrate* spiritual truth rather than spreading it—like the Sufis did in the era of Hasan of Basra and Rabi'a al-Adawiyya, or like those contemplatives who rejected 'Imperial Christianity' during and after the reign of Constantine. And the expansion of traditional esoterism into worldly dissipation not only calls for but actually *initiates* this secret countermove. When we break our identification with the expansive worldliness into which certain aspects of traditional esoterism have degenerated, we may suffer a sense of contraction—a contraction that can immediately be placed in the service of concentration. (I've used the word 'we', but the truth is that only individuals, not groups, can follow this path)

I've thought of three ways in which this concentration might express itself, which are also three supports for the spiritual Path:

(1) The re-dedication to and deepening of spiritual practice, particularly the invocation of the Name of God.

(2) The cultivation of two of three spiritual friends of like mind and spirit, so that—God willing—our sense of isolation (for those of us who are not yet content to be alone with God) will not lead to spiritual despair.

(3) The practice of shifting from the intellectual to the existential pole, from Knowledge to Being.

This third element needs some explanation. In essence, Knowledge and Being are two sides of the same Divine Reality, the *chit* and *sat* of *sat-chit-ananda*. But since the forms of traditional esoterism that emphasize *jñana* or *gnosis*, at least in their outer expressions, are designed by and for spiritual intellectuals, they tend to produce an imbalance of Knowledge as against Being, an imbalance which is possible in the realm of cosmic manifestation, though not in the realm of metaphysical principle. And this is particularly evident when the collectives dedicated to the expression of traditional esoterism begin to degenerate. We fill ourselves with spiritual knowledge, partly given by God, partly acquired through study, until we Know more than we Are. And to Know more than you Are is to be in a state where Knowledge is in debt to Being. When we accumulate Knowledge but shy away from its full realization, we are egotistically identifying with Knowledge; we are refusing to pay to Being what we owe It. (It is also possible for Being to be in debt to Knowledge—for us to refuse, due to a different form of ego-identification, to allow virtue and contemplative concentration to flower into intellectual realization; but this is not the particular imbalance I am dealing with here.)

We tend to believe that Knowledge is to be actualized only through its *application*. We know, in theory, how to do something, either in terms of worldly endeavor or of the requirements of the spiritual Path, and then we actually do it. But Knowledge can only be partially actualized through application; if we try to practically apply, in cosmic rather than metaphysical terms, our entire store of Knowledge, then we have embarked on the path of sure and certain dissipation. Certain applications of Knowledge are necessary to both worldly life and the spiritual Path, but Knowledge can only be *fully* actualized by returning to the Divine Source Who gave it, by allowing it to disappear into the spiritual Heart.

When spiritual Knowledge reaches its furthest point of expansion and manifestation, it either transforms itself into worldly knowledge and so dissipates, or it makes the great turn, the great *metanoia*, and begins its journey back to God: *To Allah does the whole matter revert.* Knowledge is a kind of substance; when the time comes for Knowledge to transform itself into Being, the many particular forms that appear inside it melt and merge; they *deliquesce* into a single *elixir*, which flows backwards through the spiritual Heart, and disappears into the Heart of God. This is how *knowledge* becomes *gnosis*.

In eschatological terms, the perfection and end of Knowledge as a separate cosmic dimension—or rather the shift in the cosmic balance from Knowledge to Being—is in line with the cyclical shift, presented by René Guénon in *The Reign of Quantity and the Signs of the Times*, from the Essential pole to the Substantial pole. Early in the cycle, Being manifests in terms of Essence or Form, of quality rather than quantity. Later on, this formal manifestation begins to become abstract, merely 'intellectual' in the colloquial sense of that term. And after a certain point, the shortest and truest way back to Being, the way most in line with the Tao, with the Will of God, is through the Substantial pole—Kali—the Black Virgin. The Essential pole is not rejected; rather, it is *invoked*, as the polar opposite of Substance, precisely by the full actualization of Substance—by the divestment of the *atman*, the Absolute Divine Witness within us, of all the forms of Its Self-expression by Mahamaya, by universal manifestation appearing as the pole of pure Substance, by Kali Ma—after which, in the dawn of pure Being, manifesting again as the Essential pole, Essence and Substance are once more known as One.

This turn has nothing to do with anti-intellectualism or the rejection of Knowledge as such. Many people, having failed to make intellectual sense of the metaphysical order and the requirements of the spiritual life, pridefully reject knowledge just as they once pridefully sought it out. But what we are talking about here is the perfection of Knowledge, not its rejection. To identify with Knowledge, to hold on to both acquired Knowledge and the crystallized forms left behind by the influx of God-given Knowledge, is to petrify. And to reject Knowledge in the name of experience, either worldly or spiritual, in an attempt to overcome this petrification, is simply to dissolve. Experience is never fulfilled on its own plane. It is the destiny of both spiritual and worldly experience, if it is to reach perfection, to be transformed into Knowledge, just as it is the destiny of Knowledge to be transformed into Being. Experience in itself is not Being, only potential Being; experience is transformed into actualized Being only through Knowledge.

And the only way for Knowledge to be perfected is for it to disappear into the spiritual Heart. This, not the stupid and prideful rejection of Knowledge in the name of devotion or self-mortification—the spiritual *voluntarism* so definitively criticized by Frithjof Schuon (who nonetheless allowed a place for it in the spiritual life as necessary to certain temperaments)—is the true *sacrificium intellectus*. We do not choose a defective animal to sacrifice because we want to get rid of it; we seek a *pure* victim, the best of its species, in order to sacrifice it to, and for, something better—something *infinitely* better.

It is time to Be what we Know: the era of the mass popular transmission of metaphysical and esoteric doctrines has now given way to the era of the hidden saints.

PART SIX

Lao Tzu's Journey

Lao Tzu's Journey

The greatness of the earth is in its humility. Where did the universe come from? Who can say. It's as if things just fell into place. By following the way things go, by not fighting gravity, everything attains its proper shape. What ripens, decays; what decays serves new life, in humility, like water. Water has no conscious agenda. It always flows down hill, but human beings seem to have other ideas. They forget that the universe obeys the will of heaven by falling, without effort, along the invisible contours of things; that birth and death are equally a downhill flow, a letting go; and that whatever flows downhill always returns to its Source. You can pretend to fight against it, and the consequences will be quite convincing—but in reality, all your fighting against It is nothing but the perfect expression of It; the only problem is, you no longer know this: and this is the essence of the human disaster. If it were not for the human disaster, the Way would only be, and flow. Since there is such a thing as the human disaster, the Way can also teach.

And so there came an old man named Lao Tzu, whose name means 'old man'. They say he was born old. Did he then become younger? Who can say. Whatever returns to its Source is younger than a newborn child, and older than the hills.

One day Lao Tzu was walking on the earth, because there was nowhere else for him to walk (and that was fine with him), When he ran into an old woman digging wild roots from the earth on which Lao Tzu was walking.

'Good morning, Old Woman,' said Lao Tzu, 'how have you been?'

'About like always,' said the old woman, 'except for all these people moving in. They have big ideas. They want to improve my standard of living. They walk all over me with their horses and their automobiles and their big boots. It's getting hard for me to collect my roots.'

'Where do these people live, Old Woman?' asked Lao Tzu.

'Oh, down the road aways, in that camp, and in that fortress, and by that river, and in that town.'

'They say there's nothing new under the sun,' said Lao Tzu; 'I think I'll travel on and see what I can learn of this wondrous development.'

So Lao Tzu walked on till he came to a forest, and in a clearing in that forest he encountered a camp of soldiers—or were they bandits? They were either very well-organized bandits or very independent soldiers. Their leader, a proud warrior, saw him, and strutted forward to confront him:

'You're Lao Tzu the Taoist, aren't you?' he asserted, without waiting for an answer. 'As a practical man, I understand the Tao as well as any. Let me tell you my view of the matter:

'All things have a goal, and struggle to attain it. That's their nature. The sapling wants to grow, if it has to split the rock to do it. The rock wants to remain as it is, even if it has to smother the sapling. It is in the nature of the tiger to bring down the deer; it wants to taste warm blood and grow strong. The deer wants to go on living, even if the tiger has to starve.

'Water flows down hill because it wants to cut a channel for itself; the rocky ground wants to prevent this.

'The Sun is like a proud invader, and the stars like a popular militia. Every morning the Sun conquers the stars; then he grows fat and complacent, till by evening he has reached old age; that's when the little people, seeing his weakness, revolt against him; they conquer him, and divide up his kingdom among themselves.

'So you see, all things are at odds. Thus the Tao may be understood as universal warfare. It is the Tao of the strong to conquer; the Tao of the weak is to be destroyed. In this way all things obey their original nature. That's what I say; what do you say?'

'It seems to me that the root of all war is fear. If rain were to refuse to fall for fear of disappearing into the earth, being drunk by animals and absorbed by roots, all things would wither away. If rocks refused to rest in place for fear of being overturned by floods and earthquakes, the earth would lose its stability. Likewise if the Sun did not rise for fear of sunset, nor the stars come out for fear of the dawn, the pattern of heaven would be disturbed.

'Fear is the principle of conflict. Whoever seeks a goal outside himself must live and die in terror. At his coming he rejoices, and all things tremble; at his death he cries out against the injustice of heaven and earth, and all things rejoice.

'Whoever does not realize that he carries his goal within himself, but seeks it elsewhere, among the ten-thousand things, is in rebellion against the Tao. In fighting against the way of things, he perfectly illustrates the way of things. Having failed to learn his lesson, he eventually becomes a lesson for others.

'Whatever rebels against the Tao does not last long. Whatever strives against the way of Heaven, Heaven grinds down, like a lathe grinding a piece of wood.'

At this the proud warrior, hiding a look of uncertainty, turned his back on Lao Tzu, and disappeared among the soldiers of the camp.

The Old Man walked on till he came to a hill, on which stood a magnificent castle. The porter of the castle called to him from a tower above the gate: 'You are walking in the precincts of the lord of this region; state your name and your business!'

'My name is Lao Tzu, and my business is following the nature of things. I have read somewhere that it is the nature of lords to extend hospitality to travelers, and entertain wandering scholars. Would this be true of your master?'

The porter disappeared for a few minutes, then stuck his head out of the tower window and cried: 'I have presented your petition before my lord, and he has graciously granted it! Prepare to enter his presence.' The castle gate opened, and Lao Tzu walked through it. Servants conducted him to the audience chamber, where the lord of the castle sat on a high seat.

'Welcome, scholar, to my presence. It is the custom of lords and kings to patronize the wise and receive instruction from them— reserving, of course, their full rights of sovereignty. By reputation you are a scholar of the Tao. If I am my people's father, the Tao is their grandfather. If they prostrate to me, I in turn prostrate to It. As It rules me, I rule them. What have you discovered of its nature through your painstaking researches?'

'I have discovered, my lord,' said Lao Tzu, 'that according to the fool, the Tao rules heaven and earth through its glory and magnificence; according to the mediocre man, it rules through wisdom and foresight; according to the true man, it rules through poverty and nothingness.

'Glory and honor are like a man standing on tip-toe; even a breeze can knock him over. Wisdom and calculation are like a man counting coins and arranging them in separate piles while bandits are plundering his house. Poverty and nothingness are like a river bed. The force of water can destroy bridges and over-turn walls, nothing can stand in its way. But the river bed, put-ting forth no effort, rules the force of that water. It contains it, shapes it, and conducts it to its Source.

'The kings of old knew the people as water, and acted as chan-nels for them. Destructive floods ceased, and agriculture flour-ished. This, sire, is all I know of the matter.'

When Lao Tzu had finished speaking, the lord of the castle bowed his head, as if deep in thought. After a while he raised his eyes, and smiled on Lao Tzu. 'Nobility has its pleasures,' he said, 'and the greatest of these is generosity. Look on the rich objects that fill this room, and choose whatever you want.'

'If that is your pleasure,' Lao Tzu replied, 'then I choose that bowl of rice and that jug of wine, since I am as much in need of nourishment for my journey as you are in need of the power to provide it. As for the wine, it is a reminder that a joyous mood, the most fleeting of all things, can only be shared among equals.'

So the lord of the castle and Lao Tzu shared a parting cup, after which the lord dismissed him, and he continued on his way.

After a while, Lao Tzu came to a village, and on the outskirts of the village, by the side of a stream, he ran into a group of women. They were laughing and talking, washing out the village laundry. Every time they pulled a new piece of clothing out of the basket, they had something to say about the owner of it. This one was a sexually promiscuous girl who would probably get pregnant and end by committing suicide. That one was a handsome devil who knew whenever the husband of a pretty wife was away at market. This one was a cunning woman and nobody's fool. That one was a miser who was so afraid to touch his money that he would probably lose it all.

'Good afternoon, ladies,' said Lao Tzu. 'You seem to be experts on the affairs of this village. Since the person who knows one thing thoroughly is close to wisdom, maybe you can help me. All my life I have desired nothing more than to understand the Tao. Perhaps you can enlighten me on this subject.'

'The women looked at each other and giggled. Then the leading woman got to her feet, put her hand on her hips, and took a step forward.

'To me, the Tao is knowing what's going to happen. It's easy to know the future if you keep your eyes open, because things never really change. Somebody who's a fool will always be a fool. A mean person will always rub you the wrong way. A spendthrift will always be poor. It's being poor that makes him want to spend money, just as angry words from the people he irritates are what makes a mean person mean. And the more mistakes a foolish person makes, the more he despairs of ever being right. So he just gets comfortable with his foolishness; it's the easiest way.

Things go on in the grooves they've worn by going. The Tao is just the way things go.'

'I imagine then,' replied Lao Tzu, 'that you ladies must be good at predicting birthmarks. And surely hail or flood have never destroyed the crops in this village, the day before harvest was to begin. But what puzzles me is why you haven't all moved to the city and made your living by gambling. It's an easy life, and I'm sure you all would be rich by now.

'The mind of Heaven is subtle, mysterious! The only way to know it is to move with the present. An insect that lives for a day cannot know the shape of the Great Year. Yet it knows what it must do in the short day of its life, even though it has no memory on which to pattern its behavior, and no way, without memory, to anticipate the day's end.

'Men, on the other hand, have a memory, which is why they can draw wisdom from the deeds of former times. But memory can fool us too. It makes a pattern out of experience so convincing that we think we must have anticipated it, when really we are only remembering it. The beggar came on a Tuesday, on Friday the milk soured, so milk always sours on Fridays, and Tuesday is beggar-day. The Tao will have nothing to do with such nets and traps! It is the finest of all things, it slips through the cracks of anticipation and memory.

'The Tao is known through blending with its present shape and motion. You can't store up your breaths for next Tuesday, nor will the breaths you took last Friday do you any good today. To limit the Tao to fortune-telling is like trying to carry the ocean away in a teacup. It cannot be done.'

At this the women looked sideways and blushed, and began criticizing and accusing each other. Lao Tzu walked on.

Near sundown Lao Tzu came to a large city. As he passed through its streets, he came upon a sordid tavern where discharged soldiers, runaway farm girls, dissipated aristocratic youths and loose city women were drinking and carousing.

'Lao Tzu!' they all cried, as soon as they saw him. 'Our brother! Come and drink with us. We have all studied your doctrines since we were babes in arms. Let things loose! Let things go!

Move with the present impulse. Become one with the great Formless. Heaven and earth don't act through calculation. They never ask about the consequences of things; it's as if they're driving blind. If not even our first parents knew what they were doing, then how can we? The Tao is like a runaway carriage, like a forest fire started by an idiot. It throws the stars like dice on a platter, then forgets to read the numbers. Join us!'

'My friends,' Lao Tzu replied, 'You are perfect illustrations of the activity of the Tao, just like the woman who eats poisonous mushrooms, or the man who sells his farm to buy agricultural equipment. The Tao has always known how to deal fairly with people such as yourselves. I would love to join you in your interesting pursuits, but regretfully I must hurry on to a previous engagement. The sun is about to set.'

Leaving the city behind and walking toward the west, Lao Tzu met the old woman again, sitting by the side of the road. She was scrubbing the roots she had collected, then putting them away, one by one, in her bag.

'Lao Tzu, Old Man! It's good to see you again. Did you learn anything on your journey through the human world?

'That's hard to say, Old Woman. You can't really learn a thing unless you somehow already know it. At times I think that learning things is like taking one suit of clothes out of your trunk while putting another away. Look east and everything to the west disappears. Look west, and the east is gone. As knowledge accumulates, the Tao is obscured. But I suppose I learned that balance is always lost, and always restored, only to be lost all over again. This is the Tao of the human world, and I always seem to be there to watch the changes.

'But if a top-heavy load falls off a wagon and rests on the ground, is that loss, or is it restoration? The Way itself is far beyond loss and restoration. Balance does not describe it.

'The only way we can ride the Tao beyond ignorance is to ride it beyond knowledge too; and whether or not we ride it, whether we stay on its back or get bucked off, it remains perfectly what it is, and we remain as perfect examples of it.'

'The Tao in itself is neither obvious nor mysterious; it simply is what it is. The real mystery of the Tao is that it can be used. By refusing to harness it, by bridling anything in us that thinks it can depart from the Way of things, we turn it into a powerful steed, a horse that can carry us far without flagging—not because we have broken it, but because it has broken the restless thing in us that's always looking for bigger and better horses.

'But as for what the Tao is in itself, all we can say for sure is that Way is the sort of load that words cannot carry. Yet we keep on talking about it, don't we? Like that Chinese poet once said, "If those who talk don't know and those who know don't talk, then where did all those books on Taoism come from?"'

'That's alright,' said the Old Woman. 'If the pot is on at a good boil, you'll have to expect some steam.'

Then the Old Woman bent over and rolled on the ground, and when she stood up again she had become a water buffalo. Lao Tzu mounted her and rode west, till they came to the border of the human world. There they met the gatekeeper, but he refused to open the gate:
'No one is allowed past this point until they've paid the toll.'
'What is the toll?' asked Lao Tzu.
'The toll is your thoughts,' said the gatekeeper. 'No one is allowed to leave the human world until they've left their thoughts here with me.'
'Alright,' said Lao Tzu. So he dismounted, and retired to the gatekeeper's cabin. On paper kindly provided by the official he wrote down all his thoughts. It took three days.
'Will this do?' said Lao Tzu when he was done.
'Perfectly,' said the gatekeeper. And he opened the gate.
So Lao Tzu mounted the water buffalo, waved goodbye to the gatekeeper, and rode west, beyond the human world.

Rain comes from the East; it's an event of the morning. The river flows toward the West, till it reaches the evening. There are those who, like salmon, have to fight upstream, and die.

They are the great spiritual warriors. But Lao Tzu will have none of this. No blame; it's simply a matter of taste. It's the way of his nature to follow the stream to the West. Long life, short life, what's the difference? Why create turbulence? Why not just float, like a leaf on the stream? Like any other man, he must work and struggle to maintain his life. But if all his works are forgotten, what does it matter? That's fine with him. His skill is simply to stay with it. That way nothing is lost. It's like the Old Woman always says:

'The valley spirit never dies.'

www.ingramcontent.com/pod-product-compliance
Lightning Source LLC
Chambersburg PA
CBHW031241090426
42742CB00007B/265